*Burkina Faso*

NATIONS OF THE MODERN WORLD: AFRICA
Larry W. Bowman, *Series Editor*

*Burkina Faso: Unsteady Statehood in West Africa,* Pierre Englebert

*Senegal: An African Nation Between Islam and the West,*
Second Edition, Sheldon Gellar

*Uganda: Tarnished Pearl of Africa,* Thomas P. Ofcansky

*Cape Verde: Crioulo Colony to Independent Nation,* Richard A. Lobban, Jr.

*Madagascar: Conflicts of Authority in the Great Island,* Philip M. Allen

*Kenya: The Quest for Prosperity,* Second Edition,
Norman Miller and Rodger Yeager

*Zaire: Continuity and Political Change in an Oppressive State,* Winsome J. Leslie

*Gabon: Beyond the Colonial Legacy,* James F. Barnes

*Guinea-Bissau: Power, Conflict, and Renewal in a West African Nation,*
Joshua B. Forrest

*Namibia: The Nation After Independence,*
Donald L. Sparks and December Green

*Zimbabwe: The Terrain of Contradictory Development,* Christine Sylvester

*Mauritius: Democracy and Development in the Indian Ocean,* Larry W. Bowman

*Niger: Personal Rule and Survival in the Sahel,* Robert B. Charlick

*Equatorial Guinea: Colonialism, State Terror, and the Search for Stability,*
Ibrahim K. Sundiata

*Mali: A Search for Direction,* Pascal James Imperato

*Tanzania: An African Experiment,* Second Edition, Rodger Yeager

*São Tomé and Príncipe: From Plantation Colony to Microstate,*
Tony Hodges and Malyn Newitt

*Zambia: Between Two Worlds,* Marcia M. Burdette

*Mozambique: From Colonialism to Revolution, 1900–1982,*
Allen Isaacman and Barbara Isaacman

# BURKINA FASO
## Unsteady Statehood
## in West Africa

## PIERRE ENGLEBERT

WestviewPress
*A Division of* HarperCollins*Publishers*

*Nations of the Modern World: Africa*

All rights reserved. Printed in the United States of America. No part of this publication may be reproduced or transmitted in any form or by any means, electronic or mechanical, including photocopy, recording, or any information storage and retrieval system, without permission in writing from the publisher.

Copyright © 1996 by Westview Press, A Division of HarperCollins Publishers, Inc.

Published in 1996 in the United States of America by Westview Press, 5500 Central Avenue, Boulder, Colorado 80301-2877, and in the United Kingdom by Westview Press, 12 Hid's Copse Road, Cumnor Hill, Oxford OX2 9JJ

Library of Congress Cataloging-in-Publication Data
Englebert, Pierre, 1962–
  Burkina Faso : unsteady statehood in West Africa / Pierre Englebert.
    p.  cm. — (Nations of the modern world. Africa)
  Includes bibliographical references and index.
  ISBN 0-8133-8249-1
  1. Burkina Faso—Politics and government.  2. Nationalism—Burkina Faso—History—20th century.  3. National state—History—20th century.  I. Title.  II. Series.
DT555.59.E54  1996
966.25—dc20                                                    96–2700
                                                                               CIP

The paper used in this publication meets the requirements of the American National Standard for Permanence of Paper for Printed Library Materials Z39.48-1984.

10    9    8    7    6    5    4    3    2    1

To my wife, Beth,
and in memory of my father, Marcel,
and my brother, Luc,
with love and gratitude to each

# Contents

List of Illustrations     xiii
Acknowledgments     xv
List of Acronyms and Abbreviations     xvii

## 1 INTRODUCTION     1

Notes, 7

## 2 "E PLURIBUS UNUM?" PRECOLONIAL AND COLONIAL HISTORY     9

The Mossi Kingdoms and the Experience of Statehood, 10
Ethnic Diversity and the Question of Common Ground, 16
A Late Colonization, 18
Settlement and Resistance, 20
The First Colony of Upper Volta and Its Partition, 23
France's Fourth Republic and the Neutralization of Voltaic Radicalism, 27
Parties and Persons: Emergence of a Pattern, 30
Reluctantly Moving Toward Independence, 32
Notes, 36

## 3 POLITICAL INSTABILITY AND THE QUEST FOR LEGITIMACY     43

The First Republic, 43
Sangoulé Lamizana: The Benevolent Dictator, 46
The Comité Militaire pour le Redressement et le Progrès National, 52
The Conseil de Salut du Peuple: A Prelude to the Revolution, 53

The 1983 Coup and the Conseil National de la Révolution, 55
The "Rectification" and the Front Populaire, 61
The Fourth Republic: An Iron Fist in a Velvet Glove, 66
The Roots of Instability, 69
Notes, 72

## 4 THE ECONOMY OF GROWTH AMID POVERTY 77

The Colonial Economy, 79
Culture and Economic Development, 80
Agriculture as Livelihood and Constraint, 83
Mining and Industrial Underdevelopment, 96
Services, 100
Macroeconomic Performance and Policies, 102
Coping with Poverty: Migration, 110
Notes, 112

## 5 SOCIETY AND CULTURE 117

Population and Demography, 117
Ethnicity and Language, 119
Religion, 127
Sick and Illiterate: The Status of Health and Education, 133
Women and the "Other Half of the Sky," 137
The Culture of the State, 141
Notes, 144

## 6 FOREIGN RELATIONS, OR THE LIMITS OF SOVEREIGNTY 149

A Small Country with a Limited Role (1960–1983), 150
A Small Country with Large Ambitions (1983–1987), 151
International Dimensions of the "Rectification" (1987–1994), 156
Pride and Passion: In the Grip of France, 160
The Enduring Libyan Connection, 162
Notes, 164

## 7 CONCLUSION: A HISTORICAL PARENTHESIS OR THE FOUNDATIONS OF ENDURING STATEHOOD? 167

A Relative Success Story, 168
Lingering Political, Ethnic, and Economic Threats, 170
Toward Steady Statehood and Sustainable Development? 172

*Bibliography* 175
*About the Book and Author* 187
*Index* 189

# *Illustrations*

## *Figure*
4.1  Wages and prices in Burkina                                         94

## *Maps*
1.1  Burkina Faso in West Africa                                          3
1.2  Burkina Faso: Vegetation                                             4
1.3  Burkina Faso: Climate, temperature, precipitation                    5

2.1  The partition of Upper Volta, 1933–1947                             26

3.1  Burkina Faso: General political map                                 44

4.1  Burkina Faso: Economy                                               78

5.1  Burkina Faso: Ethnic map                                           120

## *Tables*
3.1  Abstention in Burkina's electoral consultations                     71

4.1  Industrial origin of gross domestic product                         84
4.2  Cereal output and commercial production of main cash crops          86
4.3  Evolution of agricultural producer prices by political regime       95
4.4  Burkina, Sahel, West Africa, and Africa's
     economic performances in the 1980s                                 104

## *Photographs*
Sahel landscape in the region of Dori                                     2
A typical Mossi village                                                  13
Workers on the Abidjan-Ouagadougou railway, October 1954                 24
Mogho Naba Saaga II, August 1950                                         29
President Sangoulé Lamizana                                              47
President Thomas Sankara (with walkie-talkie) and bodyguards             57
President Blaise Compaoré during the 1991 presidential campaign          67
Songhai granary for the storage of millet and sorghum, near Dori         85

| | |
|---|---:|
| Gold field in the Aribinda region of the Sahel | 97 |
| A Dagara woman selling pottery at the market in the city of Dano | 100 |
| A Wunie (Ko) dwelling in the Boromo region | 118 |
| A Wunie (Ko) mask in the Boromo region | 122 |
| The mosque of Safane | 128 |
| President Compaoré welcomes Pope John Paul II to Ouagadougou | 132 |
| Samo dancers decorated with cowries in Kiembara | 142 |
| A still from *Weend Kuni*, a film by Gaston Kaboré | 143 |
| Human pyramid on a motorbike | 169 |

# *Acknowledgments*

This book would not have been written if it were not for Professor Paule Bouvier from the Université Libre de Bruxelles (ULB), who first introduced me to African and development studies and offered me her continuing friendship and support thereafter. I am also most grateful to my friend Eric Philippart, professor and associate director of ULB's Center for the Study of International and Strategic Relations (CERIS), who has helped me in many ways over the years. Additional thanks go to the following people and institutions who in one way or another helped me research and write this book: Drissa Aouba, the Association Burkinabè de Sociologie, Ernest Bayala, Deborah Bloch, Burkina's Centre National de la Recherche Scientifique et Technique (CNRST), Mike Chapman, Howard Dick, Denizhan Eröcal, Liam Humphreys, Pascale Joassart, Marie-Jeanne Kanyala, Andrew Manley, Rasmane Ouédraogo, Jean-Bernard Ouédraogo, and Wally Struys. None of them, however, bears responsibility for the imperfections of the final product. I received a travel grant from the Belgian Fonds National de la Recherche Scientifique (FNRS). The maps were drawn by Tam Huu Do at the Department of Geography, California State University at Long Beach.

I also wish to acknowledge my academic and professional indebtedness to Larry Bowman, Mike Chapman, Chris Craemer, Jacqueline Damon, Marion Doro, John Elliott, Gregory Kronsten, Graham Matthews, Faride Motamedi, Jeffrey Nugent, Judy O'Connor, Michael Schatzberg, Lyn Squire, Jacques van der Gaag, Arthur Wayne, and William Zartman.

My deepest gratitude goes to the three persons to whom this book is dedicated. My wife, Beth, patiently put up for many months with a husband who disappeared to Africa each time he sat at his desk. In our discussions her experience and understanding of West Africa, deeper and more sensible than mine, were a constant source of learning and motivation. Outside of work she has stood by me through the hardest times. My father, who passed away much too young in 1990, was born and spent his youth in Africa. His enthusiasm for all things and the quality of his person made me curious about the continent that had helped shape him, and his life awoke my awareness of the human dimensions of colonialism and independence. Finally, I think as always of my late brother, Luc, a lieutenant and fighter pilot in the Belgian Air Force, who was coldly grounded by a ruthless cancer in August 1994 at the age of twenty-three and whose courage and grace in the face of suffering and death will never cease to amaze and inspire me.

*Pierre Englebert*
*Los Angeles, California*

# Acronyms and Abbreviations

| | |
|---|---|
| ADES | Alliance pour la Démocratie et l'Emancipation Sociale (Alliance for Democracy and Social Emancipation) |
| ADF | Alliance pour la Démocratie et la Fédération (Alliance for Democracy and Federation) |
| ADP | Assemblée des Députés du Peuple (Assembly of People's Deputies) |
| AOF | Afrique Occidentale Française (French Occidental Africa) |
| ARDC | Alliance pour le Respect et la Défense de la Constitution (Alliance for the Respect and Defense of the Constitution) |
| BALIB-B | Banque Arabe-Libyenne du Burkina (Arab-Libyan Bank of Burkina) |
| BCEAO | Banque Centrale des Etats d'Afrique de l'Ouest (Central Bank of West African States) |
| BIAO | Banque Internationale de l'Afrique de l'Ouest (International Bank of West Africa) |
| BIB | Banque Internationale du Burkina (International Bank of Burkina) |
| BICIA-B | Banque Internationale pour le Commerce, l'Industrie, et l'Agriculture du Burkina (International Bank for Commerce, Industry, and Agriculture of Burkina) |
| BND | Banque Nationale du Développement (National Development Bank) |
| BNP | Banque Nationale de Paris (Paris National Bank) |
| Brakina | Brasseries du Burkina (Breweries of Burkina) |
| CATC | Confédération Africaine des Travailleurs Croyants (African Confederation of Workers of Faith) |
| CBMP | Comptoir Burkinabè des Métaux Précieux (Office of Precious Metals of Burkina) |
| CDR | Comité de Défense de la Révolution (Revolutionary Defense Committee) |
| CEAO | Communauté Economique d'Afrique de l'Ouest (Economic Community of West Africa) |
| CFA | Communauté Financière Africaine (African Financial Community) |

| | |
|---|---|
| CFD | Coordination des Forces Démocratiques (Confederation of Democratic Forces) |
| CFDT | Compagnie Française pour le Développement des Fibres Textiles (French Company for the Development of Textile Fibers) |
| CFTC | Confédération Française des Travailleurs Chrétiens (French Confederation of Christian Workers) |
| CISL | Confédération Internationale des Syndicats Libres (International Confederation of Free Trade Unions) |
| CMHV | Communauté Musulmane de Haute-Volta (Muslim Community of Upper Volta) |
| CMRPN | Comité Militaire pour le Redressement et le Progrès National (Military Committee for National Recovery and Progress) |
| CNCA | Caisse Nationale de Crédit Agricole (National Savings for Agricultural Credit) |
| CNPP | Convention Nationale des Patriotes Progressistes (National Conference of Progressive Patriots) |
| CNR | Conseil National de la Révolution (National Council of the Revolution) |
| COMITAM | Compagnie Minière de Tambao (Mining Company of Tambao) |
| CPI | consumer price index |
| CR | Comité Révolutionnaire (Revolutionary Committee) |
| CSB | Confédération Syndicale Burkinabè (Burkinabè Trade Union Confederation) |
| CSP | Conseil de Salut du Peuple (Council of People's Salvation) |
| CSV | Confédération Syndicale Voltaïque (Voltaic Trade Union Confederation) |
| DOP | *Discours d'orientation politique (Speech of Political Orientation)* |
| ECOMOG | ECOWAS Monitoring Group |
| ECOWAS | Economic Community of West African States |
| ESAF | enhanced structural adjustment facility |
| FAO | Food and Agriculture Organization |
| FESPACO | Festival Panafricain du Cinéma de Ouagadougou (Panafrican Film Festival of Ouagadougou) |
| FIMATS | Force d'Intervention du Ministère de l'Administration Territoriale et de la Sécurité (Intervention Force of the Ministry of Territorial Administration and Security) |
| FP | Front Populaire (Popular Front) |
| FPV | Front Progressiste Voltaïque (Voltaic Progressive Front) |
| Fr | franc |
| GCB | Groupe Communiste Burkinabè (Burkinabè Communist Group) |
| GDP | gross domestic product; Groupe des Démocrates Progressistes (Progressive Democrats Group) |

| | |
|---|---|
| GDR | Groupe des Démocrates Révolutionnaires (Revolutionary Democrats Group) |
| GMB | Grands Moulins du Burkina (Great Mills of Burkina) |
| GNP | gross national product |
| GSV | Groupe Solidarité Voltaïque (Voltaic Solidarity Group) |
| ICO | Islamic Conference Organization |
| IDA | International Development Association |
| IMF | International Monetary Fund |
| INPFL | Independent National Patriotic Front of Liberia |
| IOM | Indépendants d'Outre-Mer (Independents from Overseas) |
| Lipad | Ligue Patriotique pour le Développement (Patriotic League for Development) |
| MDP | Mouvement des Démocrates Progressistes (Movement of Progressive Democrats) |
| MDV | Mouvement Démocratique Voltaïque (Voltaic Democratic Movement) |
| MLN | Mouvement de Libération Nationale (National Liberation Movement) |
| MNR | Mouvement National pour le Renouveau (National Movement for Renewal) |
| MPEA | Mouvement Populaire d'Evolution Africaine (Popular Movement for African Evolution) |
| NGO | nongovernmental organization |
| NPFL | National Patriotic Front of Liberia |
| OAU | Organization of African Unity |
| OCP | Onchocerciasis Control Program |
| ODA | Official Development Assistance |
| ODP/MT | Organisation pour la Démocratie Populaire/Mouvement du Travail (Organization for Popular Democracy/Labor Movement) |
| OECD | Organization for Economic Cooperation and Development |
| OFNACER | Office National des Céréales (National Cereal Office) |
| OMR | Organisation Militaire Révolutionnaire (Revolutionary Military Organization) |
| ORD | Organismes Régionaux de Développement (Regional Development Organizations) |
| OVSL | Organisation Voltaïque des Syndicats Libres (Voltaic Organization of Free Trade Unions) |
| PAI | Parti Africain pour l'Indépendance (African Party for Independence) |
| PCRV | Parti Communiste Révolutionnaire Voltaïque (Voltaic Revolutionary Communist Party) |

| | |
|---|---|
| PDP | Parti pour la Démocratie et le Progrès (Party for Democracy and Progress) |
| PDU | Parti Démocratique Unifié (Unified Democratic Party) |
| PDV | Parti Démocratique Voltaïque (Voltaic Democratic Party) |
| PLO | Palestine Liberation Organization |
| PNV | Parti National Voltaïque (Voltaic National Party) |
| PRA | Parti du Regroupement Africain (Party of African Regrouping) |
| PRL | Parti Républicain pour la Liberté (Republican Party for Liberty) |
| PSD | Parti Social Démocrate (Social Democratic Party) |
| PSEMA | Parti Social pour l'Emancipation des Masses Africaines (Social Party for the Emancipation of African Masses) |
| PTB | Parti du Travail Burkinabè (Burkinabè Labor Party) |
| RAN | Régie Abidjan Niger (Abidjan Niger Railway) |
| RDA | Rassemblement Démocratique Africain (African Democratic Rally) |
| RIA | Régiment Inter-Armes d'Appui (Inter-arm Support Regiment) |
| RPF | Rassemblement du Peuple Français (Rally of the French People) |
| SAF | structural adjustment facility |
| SCFB | Société des Chemins de Fer du Burkina (Railroad Company of Burkina) |
| SMG | Société des Mines de Guiro (Mining Company of Guiro) |
| SMIG | salaire minimum interprofessionel guaranti (minimum interprofessional guaranteed salary) |
| SNEAHV | Syndicat National des Enseignants Africains de Haute-Volta (National Union of African Teachers of Upper Volta) |
| SNEB | Syndicat National des Enseignants du Burkina (National Union of Teachers of Burkina) |
| Sobbra | Société Burkinabè des Brasseries (Burkinabè Company of Breweries) |
| SOFITEX | Société des Fibres Textiles (Company of Textile Fibers) |
| SOMICOB | Société Minière Coréo-Burkinabè (Korean-Burkinabè Mining Company) |
| SOREMIB | Société de Recherches et d'Exploitation Minières du Burkina (Company for Mining Research and Exploitation of Burkina) |
| SOSUCO | Société Sucrière de la Comoé (Sugar Company of Comoé) |
| SUVESS | Syndicat Unique Voltaïque des Enseignants du Secondaire et du Supérieur (Single Voltaic Union of Instructors of Secondary and Higher Education) |
| TPR | Tribunal Populaire de la Révolution (People's Revolutionary Court) |
| UCB | Union des Communistes Burkinabè (Union of Burkinabè Communists) |

| | |
|---|---|
| UDPB | Union des Démocrates et Patriotes du Burkina (Union of Democrats and Patriots of Burkina) |
| UDSR | Union Démocratique et Socialiste de la Résistance (Democratic and Socialist Union of the Resistance) |
| UDV-RDA | Union Démocratique Voltaïque–Rassemblement Démocratique Africain (Voltaic Democratic Union–African Democratic Rally) |
| UFB | Union des Femmes du Burkina (Union of Women of Burkina) |
| UGTAN | Union Générale des Travailleurs d'Afrique Noire (General Union of Black Africa's Workers) |
| ULC | Union des Luttes Communistes (Union of Communist Struggles) |
| ULCR | Union des Luttes Communistes Reconstruite (Reconstructed Union of Communist Struggles) |
| UMOA | Union Monétaire Ouest Africaine (Monetary West African Union) |
| UN | United Nations |
| UNAB | Union Nationale des Anciens du Burkina (National Union of the Elderly of Burkina) |
| UNDD | Union Nationale pour la Défense de la Démocratie (National Union for the Defense of Democracy) |
| UNDP | United Nations Development Program |
| UNI | Union Nationale des Indépendants (National Union of Independents) |
| UNSTHV | Union Nationale des Syndicats des Travailleurs de Haute-Volta (National Union of Trade Unions of Workers of Upper Volta) |
| UPV | Union Progressiste Voltaïque (Voltaic Progressive Union) |
| UREBA | Union Révolutionnaire des Banques (Revolutionary Banks Union) |
| USTV | Union Syndicale des Travailleurs Voltaïques (Trade Union of Voltaic Workers) |
| UTA | Union des Transporteurs Aériens (Union of Air Transporters) |
| UV | Union Voltaïque (Voltaic Union) |

# 1

# INTRODUCTION

"The land of men of dignity": such is the name the leaders of Upper Volta decided their country deserved when they renamed it Burkina Faso in 1984. To many observers, it was a good fit. The new designation was an odd mix of this poor Sahelian state's three main vernacular languages. *Burkina* stands for "worthy people," "men of dignity," in Moré, the language of the main ethnic group, the Mossi. *Faso* is "house," "village," "country," or "republic" in the vocabulary of the Western Dioula.[1] As for the people of Burkina Faso, they are the Burkinabè, the suffix -bè or -nabè meaning "inhabitants" in the northern Peuls' language.[2] This confident self-baptism was meant to symbolize a departure from twenty-four years of nominal independence as Upper Volta, a period the country's young leadership perceived as years of neocolonialism under French domination. It was well in line with the views and personality of the thirty-four-year-old president, Captain Thomas Sankara, who had seized power in a coup twelve months before the renaming and might be compared to Congo's Patrice Lumumba, Guinea-Bissau's Amilcar Cabral, or Cuba's Fidel Castro of the early 1960s. Sankara put Burkina on the map by launching a revolution and waging an all-out war on scourges such as corruption, illness, illiteracy, and urban-biased economic policies. He failed, however, to pull his country out of its underdevelopment before his erratic leadership and his increasingly repressive style brought about his downfall and death in 1987.

It is a daunting task to govern a country as disadvantaged as Burkina Faso. Its leaders have had to contend with a poor geographical location; limited natural resources; a galloping population growth rate; an unstable political culture; highly politicized armed forces; recurring droughts; volatile world markets for its commodities; and a background of ethnic, cultural, and religious diversity. It comes as no surprise, therefore, that five coups have marked the transitions between six different regimes from independence from France in 1960 to the current rule of President (and former captain) Blaise Compaoré. And yet even with the military

*Sahel landscape in the region of Dori. Courtesy of the Centre National de la Recherche Scientifique et Technique.*

a virtually permanent feature of its polity, Burkina has had one of the broadest experiences with parliamentary democracy in West Africa. But despite more than fifteen years of multiparty politics, democrats have not brought Burkina more stability than did praetorians: No less than four different constitutions have provided the background to civilian regimes since 1960.

Albeit one of the poorest countries in the world, with precious few resources, Burkina has delivered an economic performance remarkable in many ways. Burkinabè have made the best use of their main asset: human labor. And when their land has become too poor to provide sufficient returns for their efforts, they have emigrated in large numbers to eke out a living in better-off neighboring countries. In addition, as unstable as they have been, Upper Volta and Burkina's successive regimes have at least usually shared a common capacity for reasonable economic policies. Burkina has thus, oddly, experienced both significant long-term growth and enduring poverty.

Burkina lies north of Côte d'Ivoire, Ghana, Togo, and Benin; west of Niger; and southeast of Mali, making it a crossroads in West Africa (see Map 1.1). Except for Ghana, all its neighbors belonged with Burkina in the Afrique Occidentale Française (AOF), the French colonial entity for West Africa, and those countries now use French as their official language. Ghana, formerly the Gold Coast, was a British colony until 1958 and is now officially an English-speaking country.

Burkina has no coastline. Access to the ocean is limited to a fast-deteriorating railroad linking the capital, Ouagadougou, with the port of Abidjan in Côte d'Ivoire and a road heading southeast toward the port of Lomé in Togo. Both are

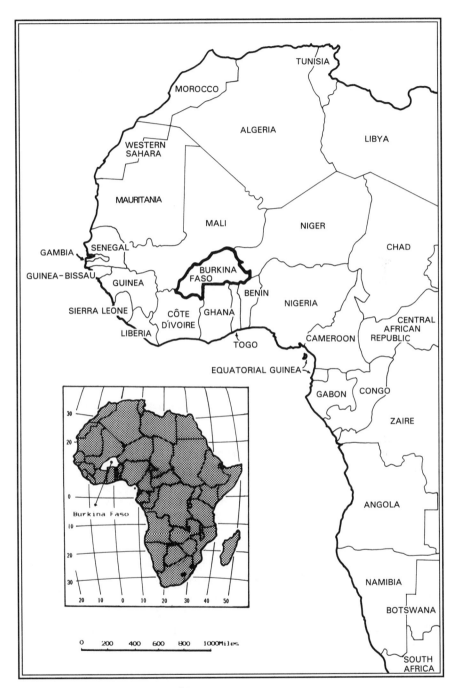

MAP 1.1  Burkina Faso in West Africa

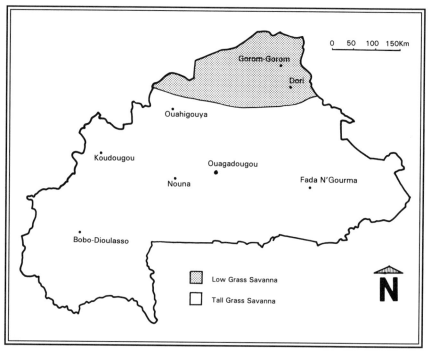

MAP 1.2  *Burkina Faso: Vegetation*

more than 1,000 km away and add substantial transportation costs to Burkina's foreign trade.

Burkina Faso is 274,200 sq km, about half the size of France and two-thirds that of California. It lies landlocked at the border between desert and tropical forest, a region referred to as the Sahel. Its north has typical Sahelian features: patches of grass and bushes amid dry and sandy plains. Its south, however, is Sudanian: greener and with denser vegetation, composing savanna and some forests. It has not yet suffered from desertification, as it is less prone to drought (see Map 1.2).

Burkina is feebly irrigated by the three Volta Rivers—the Black, the Red, and the White—which belong to the Niger River basin.[3] The Black Volta alone flows year-round. The Red and White Voltas swell during rainy season and dry up after the rains have come to an end. In a good year rains may start as early as May or June and last until September or October (see Map 1.3). During the dry season the Harmattan, a warm and dusty wind, blows from the desert. Temperatures rise to their highest level around March and April, before the rains, to peak well above 40°C.

About half of Burkina is composed of the Mossi plateau, an arid area covered with a layer of laterite that gives the earth its distinctive red color, as in many parts

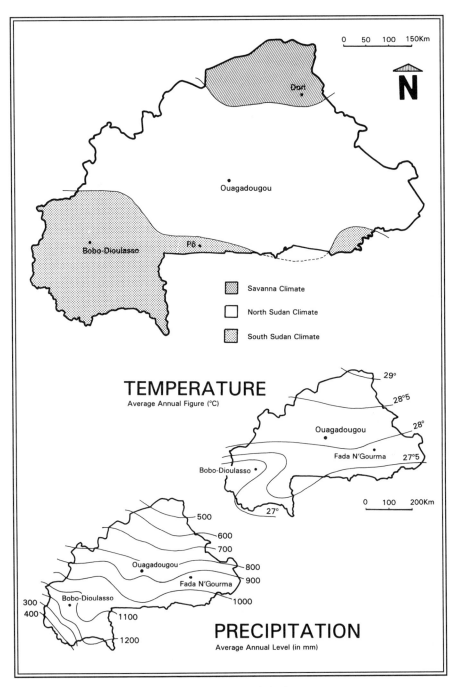

MAP 1.3  *Burkina Faso: Climate, Temperature, Precipitation*

of West Africa, but also hampers its productivity. The plateau's altitude ranges from 270 to 360 m. Landscapes are monotonous. With an average elevation of 600 m the southwest region known as the Sikasso plateau, between Mali and the Banfora cliffs, is covered mainly with sandstone.[4] At 742 m Mount Nakourour near Mali is the country's highest point.

Its social geography is as diverse as its physical milieu is monotonous. Although the Mossi, an ethnic group whose origins date back at least to the fifteenth century, account for more than half the country's 10 million people, there are about sixty other ethnic groups contributing to the country's human and cultural wealth—and potentially adding to its social instability. They are split into three main "families." The Voltaic, or Gur, family; the Mande; and the Peuls (a subgroup of the Fulani), the single representative of the West Atlantic family.

Colonization and independence brought all these relatively heterogeneous groups into a single state but failed to create a nation out of them.[5] The poor fit between the state (only recently created during the relatively short colonial period) and the institutions of preexisting societies (repressed both by the colonial and the postcolonial states) has led to a crisis of state legitimation whose consequences have been political and economic instability.

These are circumstances under which most postcolonial African states suffer, and many of the predicaments observed in Burkina to some extent can be generalized to the region. Yet, Burkina is unique in several respects. It has little ethnic conflict, civil strife, or even corruption and patrimonialism, all of which are usually common manifestations of the disjunction between state and society in Africa. Nevertheless, its political instability exceeds that of most other African countries. Together with economic volatility, this mix of civil peace and political shakiness makes for the fascinating country I portray in this book. I take as my subject Burkina as a nation, an entity, a state. If the nation has been elusive, if the entity has lacked identity, and if the state has been weak and ahistorical, these conditions are not so much obstacles to my telling Burkina's story as sources of teachings on Burkina's particular case of nationhood and statehood. In the end, it is a wonder that such an artificial creation has already endured for more than thirty years.

This book aims to be a thorough introduction. Chapter 2 highlights the historical factors of preindependence that account for Burkina's contemporary features. Its political evolution since independence is detailed in Chapter 3, which also proposes the lack of state legitimacy as an explanation for instability. Chapter 4 provides an in-depth examination of Burkina's economic situation, policies, and prospects and attempts to show their political and social dimensions. In Chapter 5 I describe the ethnic, religious, social, and cultural contours of the Burkinabè polity and in Chapter 6 explore Burkina's external relations. Looking toward the next millennium, I conclude by assessing the chances for democracy, steady statehood, and sustainable development.

## Notes

1. Because the term *Faso* contains the notion of republic or country, the country can also be referred to simply as Burkina. I will indiscriminately use both the short and long versions of the name.

2. The official spelling, as it appears in Circulaire 84-15 CNR.PRES. SGG-CM in the government publication *Journal Officiel* 803 (16 August 1984), is printed in uppercase letters (BURKINABE) so that no accent appears on the last letter. Yet it is followed by a suggested pronunciation in parentheses that reads "Bourkinabè." Subsequent official documents and publications in French by nationals have used the grave accent on the final *e*, as I did in my 1986 book, *La Révolution burkinabè*. Yet there have been many other spellings since then. Elliott Skinner, the principal U.S. authority on Burkina, writes it *Burkinabé*. Some World Bank publications use *Burkina-ba*, while the French daily *Le Monde* uses *Burkinais*, which would translate as "Burkinese." Most authors avoid the problem by using the substantive *Burkina* as an adjective (e.g., "the Burkina economy"). For lack of a more compelling choice, for the sake of consistency, and acknowledging that the ultimate say on spelling belongs to the name's inventors, I use *Burkinabè* in this book.

3. The country's former name, Upper Volta, was a reference to these three rivers, which originate in Burkina before merging and flowing south to Ghana and the Gulf of Guinea. They took their name from Portuguese explorers, *volta* meaning "return." Until 1984 the country's flag was composed of three vertical stripes—black, white, and red—representing the three Voltas. In 1984 the rivers were renamed Mouhoun (Black Volta), Nazenon (White Volta), and Nakambe (Red Volta).

4. R. J. Harrison Church, "Burkina Faso: Physical and Social Geography," *Africa South of the Sahara 1995* (London: Europa Publications, 1994), 191.

5. For a brilliant and highly readable presentation of this argument on the continental scale, see Basil Davidson, *The Black Man's Burden: Africa and the Curse of the Nation-State* (New York: Times Books, 1992).

# 2

# "E PLURIBUS UNUM?" PRECOLONIAL AND COLONIAL HISTORY

Old societies make up young Burkina. Some sixty ethnic groups, differentiated by elements of their social, economic, and political organization and at times by language and culture brought civilization and government to the region several centuries ago. One of them, however, the Mossi, soon came to dwarf all others by its demographic weight and its hegemonic ideology of statehood.[1] Its kingdoms endured from the fifteenth century to colonization in the late nineteenth and matched contemporary Burkina in terms of effective institutions and exercise of power. Alongside the large Mossi kingdoms that covered the eastern and central regions, small societies based on the household, clan, or village and often referred to as anarchical or acephalous predominated in the western part of the country. When French colonizers arrived, they imposed their own concepts of statehood and economic production, often in contrast to existing patterns. French policy gradually went from destruction to preservation of the Mossi system, and in the end occasionally allowed for an ambiguous duality of political institutions. The French often forced the stateless societies of the western regions to accept new political structures, either extending the power of neighboring rulers or establishing direct French rule.

The first sections of this chapter describe the Mossi kingdoms, their history and political and economic systems, selected non-Mossi societies and their systems, and the relations that existed between these different communities prior to colonization. I then review the relatively short colonial episode, beginning with its conquest (1888–1895), French administration and local resistance to it (1895–1945), and the progressive replacement of the French by domestic elites leading toward independence (1945–1960). I wish to stress two things: The first is the contrast between the preexisting social structures and the colonial system; the second is the creation of independent Upper Volta and its political elite as a product of colonization rather than precolonial history. In other words, when colonization ended, much of the political and economic system associated with it remained. There was no return to the political legitimacies and modes of production of the precolonial era. The new political elites were products of French education and, for all practical purposes, former collaborators with the French. The Mossi and other political systems no doubt retained a cultural value but failed to maintain their operational political significance. Herein probably lies an essential variable in accounting for the fragility and unsteadiness of the successive postindependence institutional arrangements.

## The Mossi Kingdoms and the Experience of Statehood

### *Historical Foundations*[2]

Although the Mossi, who account for more than half the population of Burkina, are the most studied of its ethnic groups, accounts of their history are still fairly imprecise.[3] Not only are there several contradictory stories about their origins but also conflicting dates for their settlement, with discrepancies as large as four centuries in accounts of the kingdoms' foundation.[4] This book follows French anthropologist Michel Izard's dating of the beginning of Mossi history around the fifteenth century, but this choice is by no means beyond debate.[5]

There are several variations on the legend of the foundation of Mossi society.[6] A. A. Dim Delobson, a Mossi himself, asserts that Niennega, the teenage daughter of the king of Gambaga (in what is now Ghana), rebelled against her father and, disguised as a man, escaped from his kingdom on a horse, taking with her a group of courtiers. When she reached what is currently Yanga, she stopped by a house that belonged to an elephant hunter named Ryallé. The latter had little choice but to welcome this imposing group of foreigners and pledge allegiance to Niennega, whom he duly mistook for a man. One day, though, she confided to Ryallé that she was a woman, "gave herself to him" (as Dim Delobson says), and eventually married him. They had a son whom they named Ouédraogo—the Stallion—in honor of the horse that had brought her there. When Ouédraogo turned fifteen, he went to see his grandfather, who lavished on him wealth, horses, and cattle. Suffering from their country's overpopulation, many locals—the Dagomba—followed Ouédraogo when he took leave of his grandfather. Back north they founded the village of Tenkodogo, which quickly grew as more Dagomba joined them and as local populations submitted to Ouédraogo. Meanwhile, the latter married a woman named Pouirketa, with whom he had several children. When his sons became adults, he made them chiefs in neighboring regions that he had brought under his influence. The eldest, Diaba Lompo, became head of the eastern Fada N'Gourma region. The second, Raoua, was sent northwest to Ouahigouya, where he founded the Zondema kingdom, which later crumbled and was replaced by the Yatenga kingdom. A third son, Zoungrana, stayed with his father.

The local Ninissi, who were at war with neighboring peoples, offered one of their daughters (named Pougtoenga "because she had a beard"[7]) to Ouédraogo, who gave her to his son Zoungrana as a wife. In exchange the Ninissi requested that Ouédraogo and Zoungrana take their country under protection. After Ouédraogo's death, Zoungrana offered the Ninissi a leader among his children; the Ninissi chose Oubry, the son of Pougtoenga, a woman of their own race. When Oubry came of age, he accordingly took over the region of his mother's birth, settling at Oubritenga ("the land of Oubry") and eventually conquering the region of Ouagadougou and Koudougou. He took the title of *mogho naba*, "king of the world," around 1495 and thereby founded the Mossi dynasty of Ouagadougou.[8]

In addition to illustrating the penetration of the Mossi from the Dagomba kingdom of northern Ghana upstream along the White Volta all the way to Ouahigouya, this account serves as an ideological foundation of the Mossi state system and its blend of conquest and assimilation, where marriages with autochthonous people play a crucial role and where power is transmitted from father to son. Yet the story is idyllic in its portrayal of indigenous populations. It is important for the Mossi as a matter of course to depict local populations as willing to submit to them, as Ryallé did with Niennega and as the Ninissi did with Zoungrana. Yet not all gladly married or warmly welcomed Mossi warriors. Among the Ninissi, Kibissi, and Gurunsi who previously occupied the land, those "who fomented revolts" and "did not want to submit" were either reduced to slavery or forced to flee.[9] Many Gurunsi sought refuge in the region of Pô and Léo, where they live to this day. The Kibissi, also known as the Dogon people, fled for the cliffs of Bandiagara in today's Mali. Many Ninissi (also called Tinguinbissi—"the first settlers") stayed, however, and their elders (the *tingsobas*) took on a sacrificial and religious role in the Mossi system.[10] The Mossi race ought thus to be understood as a combination of the Dagomba who came from current Ghana and preexisting local ethnic groups. The assimilationist feature of Mossi culture remains of paramount importance in understanding current interethnic relations.

As the historical narrative suggests, there are several Mossi kingdoms linked by a common ancestry. Primus inter pares, the Ouagadougou kingdom eclipsed all others.[11] Yet Ouagadougou should not be regarded as the capital of an alleged Mossi "empire," as there was considerable autonomy, and even infighting, among the different kingdoms and principalities. The most important rivalry was between the kingdoms of Ouagadougou and Yatenga.[12] The latter disputed the legitimacy of the Ouagadougou succession from Oubry, as the kingdom of Yatenga was founded around 1540 by Yadega, the grandson of Naba Oubry, on the ruins of Zandoma, the original local Mossi kingdom headed by Raoua. The Mossi from Yatenga gradually moved into the bend of the Niger[13] and in the sixteenth century waged wars with the Songhai, alternately sacking Timbuktu and being sacked by the Malians.

The original kingdom of Tenkodogo declined after the foundation of Ouagadougou, becoming quite marginal. As for Fada N'Gourma, it lost its political identification with the other Mossi groups and became the separate people of the Gurmanche.[14] Although Ouagadougou, Yatenga, Tenkodogo, and Gurmanche are the largest and best known, there have been as many as nineteen fairly autonomous kingdoms, including Lalgaye, Ouargaye, Konkistenga, Yako, Tema, Mane, Bousouma, Boulsa, Koupela, Bousou, Darigma, Niesega, Risiam, Zitenga, and Ratenga. The smaller principalities are autonomous but not independent from Ouagadougou. Despite their differences, however, there is an exceptional unity of political ideology and social organization among these kingdoms that warrants their study as a group.

## *Political and Social Organization*

The sociopolitical system of the Mossi is linked to the nature of its historical foundations. A distinction is made between the descendants of the conquerors and warriors, the *nakombse,* whom Izard labels "men of power," and the preexisting groups who have been assimilated, the *tengbiise,* or "men of the land."[15] The *nakombse* claim lineage to the founding fathers, Ouédraogo and Oubry. This claim is a necessary condition of political sovereignty for every Mossi chief, from the *mogho naba* down to the village chief. But that alone is not enough; the chief must also possess the *naam,* "that force of God which enables one man to control another,"[16] a concept close to the idea of the divine right or natural right of monarchs in Western political culture. The *naam* is transferred by a chief to his son. However, should the son fail to exercise it (when, for example, there are several sons in the line of succession, as would often be the case in a polygamous society), the power is lost to his descendants. Thus a large majority of the *nakombse* have lost possession of the *naam.*

The *tengbiise* do not possess the *naam* and can never acquire it, for they do not share Ouédraogo's and Oubry's ancestry. Although excluded from the exercise of temporal power, they remain Mossi in their own right, however, for they do have a compensating power. Called *tenga,* it is the spiritual power and the right to decide on matters related to land issues. There is a symbiosis between the two powers—and the two social groups that hold them—because, as Izard points out, *naam* needs the endorsement of *tenga* to be legitimate: "From the people of the land, the Moose ask the legitimation of their power because, strangers to the territory where they exert it, they can face Tenga only indirectly."[17] Without *tenga,* a new king could not "activate" his *naam* and hence materialize his power. This relationship exemplifies the extent and the two-way dimension of Mossi assimilationism and helps explain the rapid submission of indigenous peoples.[18] In

*A typical Mossi village, with family compounds, housing, granaries, and cooking utensils in the forefront. Courtesy of the Centre National de la Recherche Scientifique et Technique.*

Claudette Savonnet-Guyot's judicious words, the Mossi state was thus "born of the encounter of warriors and peasants."[19]

There is, however, considerable practical fluidity among these statuses.[20] Furthermore, the Mossi political system developed checks and balances to attenuate the powers of the *nakombse*. It is through one of these ingenious institutions that the Mossi evolved from what was essentially a lineage-based system to one that had the bureaucratic attributes of statehood. In order to defuse the potential power of the *talse*, the *nakombse* who had lost the *naam*, the king entered an alliance with them by appointing some of them *nesomba*, members of his court, a function the French colonizers understood as that of minister. The *nesomba* were individually responsible for specific domains of power and for representing specific segments of society and collectively for the designation of the king's successor upon the latter's death. Despite the established rules of succession, there remained much choice for the *nesomba* because several elder sons could be eligible, as could brothers and cousins of the king (in Yatenga) and virtually any son who could make a case for being named king. The *talse* were thus granted an essential power. In Yatenga the principal *nesomba* were the *baloum naba* (head of all the male servants, intendant of the king), the *togo naba* (spokesman), the *widi naba* (groom), and the *rassam naba* (treasurer, responsible for the king's slaves and commander of the corps d'élite in the Mossi army). There were many more lesser officials. In Ouagadougou there were five "ministers": the *widi* and *baloum nanamse*,[21] the *larale naba* (in charge of royal mausoleums), the *gounga naba* (chief of infantry), and the *kamsaogo naba* (chief of the palace eunuchs).[22] Other functions of the *nesomba* were the selection of canton chiefs among the candidates eligible through lineage and the administration of provinces. The *nesomba*'s functions, which were reproduced at every level from the *mogho* and Yatenga *nanamse* to the village chiefs, eventually became hereditary, and dynasties of *nesomba* appeared among the *talse*.[23] According to Izard, these civil servants, in addition to contributing "to the maintenance and the reproduction of the monarchical function" also marked "the presence, and the permanence of the presence, of the state."[24] The *nesomba* can thus be perceived as the state structure and apparatus of the Mossi. They override the king-to-village relationship and introduce elements of unity and bureaucracy in an otherwise decentralized and family-based system. They contribute to the cohesion of a heterogeneous society and nuance the absolutism of the power of the *mogho naba*.

In administrative practice, authority spread from the kingdom to the province, the canton, the village, and the ward, each level virtually reproducing the same power structure. The link between the *mogho naba* and the smaller principalities and villages was not one of feudalism, however, because it did not entail an exchange of loyalty for land or livestock. The land always remained the property of the *tengbiise*.[25] The Mossi kingdoms should thus not be equated with the early European state.[26] The decentralization of the system also applied to the administration of justice, which was rendered from the ward to the king in many levels of appeals and jurisdictions. There was no separation between executive and judicial powers. There was no regular army, but all adult men were liable for military service and wartime conscription.

## *Economic Organization*

The precolonial Mossi economy was based for the most part on household subsistence farming, although there was rudimentary division of labor, with blacksmiths providing tools in return for food. Beyond subsistence, two levels of exchange coexisted. Village markets held every three days supported domestic exchange between households and villages of the same region, while international trade, managed by a caste of Muslim merchants, the Yarse, took place in large cosmopolitan markets held at Ouagadougou, Bere, Dakay, La, Mane, Yako, and Koupela. The Yarse, like the blacksmiths, were neither people of power nor of the land. They were, however, an essential element in the trade route between the coastal areas and the Sahara, a route in which the Mossi kingdoms occupied a strategic central position at the crossroads of the two regions. The Mossi kingdoms exported cotton, iron tools, livestock, and slaves. From trade caravans heading north from the coast, they imported mostly kola nuts. From the caravans returning south from the Sahara, they bought salt and dry fish. Cowries and cotton bands were used as currency.

Slavery was widespread. Some people, often Gurunsi or Samo, were slaves because they had been captured or bought (the Yemdaogo). Others were descendants of slaves and thus slaves themselves (the Dapore) or former slaves of the Peuls who had left their masters to pledge allegiance to the Mossi (the Bengare). Finally, there were those who volunteered to be slaves in exchange for the protection of the *naba* (the Kamboiense). Originally used to work in the fields and for domestic aid, slaves began to be sold abroad at the beginning of the nineteenth century under Mogho Naba Naongo. Every Mossi, whether *nakombse* or *tengbiise,* could own slaves, but none could dispose of the life of a slave. Owners who killed their slaves had to pay 100,000 cowries to the local *naba*. The slave had the right to grow food, practice a craft, or raise cattle and could keep money made from such activities. Freedom could be bought for 100,000 cowries. Freed slaves sometimes remained in the village of their former masters and ended up having slaves of their own.[27]

The Mossi state derived revenues from several sources. The *mogho naba* received agricultural goods, livestock, and money from his subjects; collected heavy taxes on the sale of slaves; and had a monopoly on the sale of eunuchs.[28] Trade caravans passing through the kingdoms were also taxed and expected to present gifts to the *naba*. Yet the Mossi state does not appear ever to have derived substantial wealth from its economic activities nor did the *nanamse* and *nakombse* turn into an asset-accumulating capitalist class. By and large, the observation by French explorer Louis Binger in 1888 that the standard of living of the Mossi chiefs differed little from that of their people seems accurate. The likely reason for this was the economic role of Mossi chiefs as the "center of redistribution systems that embraced their entire territory,"[29] from the level of the village chief to that of the *mogho naba*. In the words of Elliott Skinner, a village or district chief

> was seldom wealthy, because he always had to use his revenue to fulfill unexpected obligations toward his subjects, his superiors in the political hierarchy, and his household. He had to be generous to his subjects, providing millet water, millet beer, and the more expensive imported kola nuts for those who visited and conferred with him daily. He was expected to provide food and shelter for the countless strangers who settled in his district, until they could provide for themselves. He also lodged and fed the messengers who brought him orders from the capital, and gave them gifts to take back to the provincial ministers. And he had to provide goods for the many members of his family who depended upon him. Finally, in the event of crop failure or famine in any of his villages, he had to provide grain from his granaries to feed the hungry.[30]

Izard makes a similar point with respect to the Yatenga *naba:* "The [Yatenga *naba*] has indeed more women to marry, more horses to ride, more millet to eat than his average subject, whose living standard has always been quite precarious. But ... the obligation of redistribution that burdened those bound to the responsibility of generosity did not allow them to accumulate considerable wealth."[31] The Mossi economic system thus proved better at redistributing wealth than at accumulating it, putting welfare ahead of growth. This tendency may explain the relatively poor record of development of the Mossi kingdoms at the end of the nineteenth century suggested by anecdotal evidence.

## *The Mossi at the End of Their Independence*

Binger was not impressed on his visit to Ouagadougou in 1888:

> Mossi is in a period of decadence.... The Mossi are now quite incapable of mounting expeditions like [they used to].... Mossi is now a torpid country which has allowed its neighbors to overtake it in civilization.... The Mossi are taking it easy, cultivating what they need for subsistence but no more, so that even if there are no paupers in their country, there are virtually no rich men either. Everybody just rubs along.... Inhabitants are extremely apathetic.[32]

Although Skinner disagrees with Binger and believes that the Mossi kingdoms were not decadent but simply "structurally weak," their economic base never having been strong enough to support a high degree of centralization,[33] it seems nevertheless that by the end of the nineteenth century the Mossi's golden age belonged to history. Dim Delobson notes that the country had become quite insecure in the period immediately preceding the French arrival. The *mogho naba* had been fighting with some other kingdom's *nanamse*. The Songhais, whom the *mogho naba* had enlisted to help him defeat his opponents, were ravaging parts of the country.[34] The political order was crumbling. The French saw this and used it to their advantage: The Mossi was ripe for colonization.

## Ethnic Diversity and the Question of Common Ground

The abundance of research and documentation on the Mossi should not overshadow the remarkable wealth and diversity of other ethnic and social systems in precolonial Upper Volta. Besides the Mossi and the relatively similar Gurmanche, dozens of other groups experienced their own specific precolonial histories. Some of the largest were the Gurunsi in the south; the Bwa, Bobo, Lobi, Senufo, Marka, and Samo in the west; and the Peuls in the north. With respect to their political organization, most of these ethnic groups are referred to as stateless and many as acephalous—or without the institution of chiefdom.[35]

The Birifor, for example, a branch of the Lobi group that shares with the Dagara the land along the segment of the Black Volta that marks the border with Ghana, was a society with no central authority, without even chiefs or villages. Lineage determined everything, and authority belonged to the elder within each lineage. The *yir,* a family farm consisting of a household compound with animals and a small adjacent plot of land, was the unit of political and economic organization. Everyone worked on the community fields in priority. The crops were collectivized and the harvest distributed to the women of the *yirs* every third day to prepare food for their families. Relations among *yirs* were structured by an elaborate system of checks and balances that helped avoid both excessive political and economic strength of one *yir* over the others and food shortages.³⁶

The Bwa, Samo, and Senufo were other such stateless groups.³⁷ Yet in their cases the political unit was the village. The Bwa, for example, whose 450 villages overlapped what are now Burkina and Mali, added to the principle of lineage that of the territoriality of the village in which several families were concentrated. No allegiance was granted above the village level. Each village, which could count up to 1,000 members, was a completely autonomous and self-centered political unit. Within the village, however, the first unit of economic and political identity was the family. What made the different Bwa villages one single ethnic group was a common religious belief in a single god, Do. The leadership of each village belonged to the eldest of the founding lineage, assisted by a council of elders. Within the village labor was divided among farmers, blacksmiths, and priests. There was endogamy among these different groups whose position was acquired by heredity, thereby preventing class conflicts.

Finally, the Gurunsi presented a type of village-based society in apparent transition toward statelike organization. The French penetration appears to have interrupted a process by which the authority of one village was progressively extended over several other villages, moving toward an increasingly centralized system.³⁸

This superficial sampling and survey of Burkina's social practices hints at their diversity and contrasts: societies articulated in the state, the village, and the family. The question arises—typical in the field of African studies—whether this multitude of ethnic identities can provide the basis for a nation. What they shared could provide a first clue. Skinner stresses the common characteristics of the peoples of the "Voltaic culture area,"

> all of whom speak languages belonging to the Gur or Moshi-Grunshi group of the Niger-Congo family of African languages. . . . These groups are primarily horticultural; cereals are their main crops, and they cultivate yams where conditions permit. . . . The veneration of ancestors lies at the core of their religious beliefs, but the existence of an otiose high god . . . is also recognized. There are also tutelary spirits, the chief one being a female earth deity . . . called *Tenga*. . . . The Tengsobadamba, who are priests or custodians of the earth shrines, are found among all the groups in the Voltaic family; and they often share leadership with more secular chiefs called Naba, Na, or Nab. . . . Most of the ruling clans of the Voltaic groups claim common ancestry and acknowledge the same totems.³⁹

The problem with Skinner's list of similarities is that it applies to a family of ethnic groups that extends beyond the territory of Burkina to Côte d'Ivoire, Mali, Benin, and northern Ghana and would support the claim for supranational integration better than that for nationhood. Savonnet-Guyot proceeds differently, surveying the extent to which Mossi hegemony has uniformized the other groups into a nation: The Mossi "system . . . covers a society . . . empowered by its long expansionist and assimilationist tradition to master . . . a space that today has become 'national' [and] of which it can reduce the differences and asperities."[40]

She contemplates a role for the Mossi as the cement of a nation because of their demographic domination and assimilationist outlook. The integrative nature of the Mossi indeed suggests a potential avenue for nationhood. The Mossi have colonized many groups, offered shelter to some, and married into others.[41] Rather than fully imposing their system, they have incorporated social, political, and cultural elements of the groups they have integrated in a two-way process, creating a new joint identity out of each encounter. Unlike pure colonists, the Mossi have diluted their nature in that of the occupied peoples and shaped identity out of diversity. By the end of the nineteenth century, however, this process did not extend beyond the northern, eastern, and central peoples, leaving the whole western part a sharply different set of political societies.[42]

There is thus a case to be made for a certain cultural unity of the peoples of Upper Volta, except for a strong east-west schism in terms of political culture. The case for unity, however, extends beyond Burkina to neighboring countries and, to a lesser extent, to most of sub-Saharan West Africa, and is independent of the area and shape of Burkina. In fact, it only serves to highlight the artificiality of Burkina's borders. The peoples of Burkina are not so different from one another that the centrifugal forces of culture, language, and so on would fragment and dissolve the state, yet there is no reason why they and not others should have been brought together in a state apart from the arbitrariness of the French colonial administrative partition. It is to this episode of foreign invasion, state destruction, and state re-creation that we turn in the next sections.

## A Late Colonization

Until the mid-1880s the peoples of Upper Volta remained essentially unbothered by colonial visitors. Following the 1884–1885 Berlin conference among colonial powers, however, Germany, France, and Britain competed to move north from their holdings in the Gulf of Guinea (and east from Senegal, for France) in order to be first to lay claim to the territories located in the loop of the Niger River. France's captain Binger, coming from Kong in Côte d'Ivoire, was the first person officially to represent a colonial power in Ouagadougou in June 1887. Although he failed to convince the *mogho naba* to turn his kingdom into a protectorate, he did secure such a treaty from the Ouattara leader of Kong for "all his states," which the French understood as including the whole southwestern region of today's Burkina.[43] In 1888 the French lieutenant colonel Parfait-Louis Monteil again failed to seal a protectorate deal with the *mogho naba* but signed two local treaties in the bend of the Black Volta and two more with the Peul leaders of Dori and Sebba. Thus the French had encroached on some significant parts of Burkina to the west and the north of the Mossi kingdoms.

In 1890 François Crozat, a French naval physician assigned to West Africa, was sent on a mission to Ouagadougou with presents for the *mogho naba*. According to Dim Delobson, the gifts were refused and the envoy expelled by a *naba* who was growing weary of the increasing passage of white men in his kingdom.[44] From 1892 to 1895 there followed several more unsuccessful French attempts to establish a protectorate over the Mossi from Côte d'Ivoire.

It turned out to be Great Britain (in the person of George Ekem Fergusson, the son of a Methodist minister from the Gold Coast) that first succeeded in signing a treaty with Mogho Naba Wobgo, on 1 July 1894. Though only a treaty of friendship and free trade, it nevertheless formalized the *mogho naba*'s promise "not to accept any protectorate or sign any agreement with another power without Britain's consent."[45] Yet this would remain an isolated British strike and a dead-letter commitment on the part of the *naba*. On 20 January 1895 the French commandant Decoeur, coming from Benin, signed a protectorate with the king of the Gurmanche, barely beating the Germans to it.

With the conquest of the Gurmanche, the Mossi were left surrounded by French protectorates to the east, west, and north and by the British to the south; their own colonization was but a matter of time. Infighting among Mossi kingdoms would make the colonizers' task easier. The determinant missions in their conquest came from the west. On 18 May 1895, Captain Destenaves of France signed a protectorate with Yatenga Naba Baogo, who needed his help to fight off his longtime opponent, Bakharé, who had pretensions to the throne. Destenaves died in June in a confrontation with Bakharé's forces, and the latter then successfully took over as Yatenga Naba Bulli. Yet the protectorate was not threatened, for the new Yatenga *naba* immediately requested its renewal, and the French obliged on 1 November 1895.

The French protectorate over the Peul country was formalized on 27 April 1895. On 25 September 1895 it was Bobo-Dioulasso's turn to be conquered despite its resistance. Following the fall of Yatenga, the race for Ouagadougou intensified. In 1896 French lieutenants Voulet and Chanoine were sent to beat the British to Ouagadougou and Sati, the Gurunsi capital. Warned a day ahead of time of the arrival of the French column, Naba Wobgo decided to flee Ouagadougou but sent a small army to face the French. Four shots in the air apparently sufficed to disperse the imperial army, terrified at the sight of the red *chechias,* the trademark hats of the Tirailleurs Sénégalais, the French colonial army corps. The Voulet-Chanoine column tried unsuccessfully to catch the emperor then returned to Ouagadougou and set much of the city afire.[46]

More villages would fall victim to French arson and more local battles would be waged before the French retreated to Timbuktu for lack of food and ammunition. On his way back to Ouagadougou three months later, Wobgo found out that the French, too, were returning, with reinforcements that included supporters from colonized Yatenga. The *mogho naba* aborted his return. In Wobgo's absence his half brother Kouka was made Mogho Naba Sighiri by the French, who forced the *widi naba* to confirm the appointment. Kouka signed the protectorate on 20 January 1897. In February 1897 the French troops coming from the Sudan met with those from Dahomey in the Gurmanche, effectively sealing French control over most of what was to be Burkina. A treaty with Germany in July 1897 confirmed France's rights over the Gurma, and on 22 April 1897 an agreement with the British gave France control over the western Gurunsi country.

From June to December 1897, the French tried to capture Mogho Naba Wobgo, who had sought British protection after fleeing from Ouagadougou. But in June 1898 the French and British agreed on the border between what would be the Gold Coast and Upper Volta. At the eleventh parallel north, it gave the whole Mossi country to the French. Wobgo retired with a British pension, near Gambaga in the Gold Coast, where he died in 1904. The French harshly repressed the southern Mossi who had helped him.

## Settlement and Resistance

The marginal and landlocked territory of Upper Volta did not warrant the immediate creation of a colony in the eyes of the French. The territory was first declared a military region in October 1899 and placed under direct control of the governor-general of French West Africa. In October 1902 it became part of the military region of Sénégambie-Niger, with Kayes as capital. It was finally made a colony in October 1904, when Sénégambie-Niger was replaced with Haut-Sénégal-Niger, with its own lieutenant governor and administration in Bamako.

In 1907 the colony was divided into administrative units called *cercles* with little regard for preexisting administrative and political divisions. The area east of the White Volta comprised the *cercles* of Kaya, Dori, Fada N'Gourma, and Tenkodogo. The central region was split among Ouagadougou, Koudougou, and Ouahigouya and the western part into the *cercles* of Bobo-Dioulasso, Dedougou, Gaoua, and Batié. This arrangement was modified in 1912 when all Mossi *cercles* were merged into the Cercle du Mossi. Each was headed by a *commandant* and divided into *subdivisions* and *postes*.

The French set out from the beginning to weaken the existing political systems in order to make room for theirs. Because of their relative institutional strength, the Mossi kingdoms took the most attention. After the death of Mogho Naba Sighiri in 1905, the French manipulated the appointment of a new *naba* who, because of his young age, would be more amenable to their needs: Saidou Congo, Sighiri's son who became Mogho Naba Kom II, was sixteen. Following his appointment the French began restructuring the Mossi kingdoms (like the rest of the region) into *cercles,* thereby occasionally creating situations where one district chief had power over another one and replacing the *nesomba* with individuals "more favorably disposed toward their rule."[47]

The imposition of *indigénat,* a system by which French administrative authorities had the power to imprison and impose fines on "natives," without recourse to justice or possible appeal, for such offenses as "committing any act of a nature to weaken respect for French authority" or "manifestations troubling public peace,"[48] also removed judicial powers from the hands of the chiefs. Slavery and some religious practices deemed to have a political content were abolished, further diminishing the chiefs' status. Although all chiefs were targeted, it was the *mogho naba* who lost the most autonomy. Soon all his decisions had to be authorized by the French. The Mossi political structure was decapitated, with the *cercle* commanders now heading the chain of authority that went down to the village chiefs through the "provincial ministers." In about fifteen years and facing surprisingly little resistance, the French had reduced the *mogho naba,* the heir to a 400-year-old dynasty, to a figurehead and had entirely tamed the Mossi political system.

Although they crushed the political dimensions of chiefdom, the French nevertheless extensively used its services as a colonial auxiliary. Chiefs were enjoined to collect taxes (and received a percentage thereof) and recruit men for forced labor and, later, conscription. On the ethnic groups that did not know the institution of chiefdom, the colonizers imposed chiefs, often exterior to the societies they were to administer.[49] In the *cercle* of Dedougou, for example, the French encouraged all chiefs capable of exerting their authority over previously independent villages to do so. Thus originated a new class of chiefs "generated by the colonial administration" in an attempt to replicate existing chiefdoms but deprived of any historical legitimacy and sometimes downright unpopular with local populations.[50] The Ouattara family, which ruled the region of Kong in Côte d'Ivoire, saw its responsibility extended over the *cercle* of Bobo-Dioulasso. Yet at times the system of indirect administration and local auxiliaries was not possible. In the Lobi *cercle,* where there was no authority above the head of a lineage, the French received very little collaboration and resorted mainly to direct administration and a direct levy of taxes.[51]

French colonization claimed a civilizing mission. Domestic slavery was officially banned in 1901 but was in fact tolerated by local administrators. The slave trade was criminalized in 1903 and actually ended in 1914, but slavery went on in many forms.[52] Schooling began, mostly thanks to the Pères Blancs missionaries, who opened their first school in 1901. The improvements in living standards were limited, however. In 1908 and 1914 droughts triggered famines and epidemics against which the French did little, fostering resentment and further opposition to colonial rule.[53] In addition, French administration imposed a heavy and destructive burden on the country. The main plagues that hit Africans were taxation, forced labor on infrastructure projects and in cotton fields, and military conscription. Taxation reached such punitive levels that Africans had to sell their assets and thereby transfer their wealth to the French administration, which effectively expropriated them without compensation. Failure to pay resulted in the sequestration and sale of an individual's goods by the local chief and humiliating punishments.

The expenses of the colonial administration were allocated first and foremost to improving communications. Roads, telegraphic connections, and bridges were built with forced labor. Next came the development of cash crops. Cotton, silk, and rubber were introduced without much success.

In 1912 military conscription was introduced in French West Africa, with registration of all men aged twenty to twenty-eight; by August 1914, Mossi, Gurmanche, and Gurunsi troops contributed to the defeat of Germans in Togoland. In 1918 conscription was extended to all men eighteen to thirty-five, and the following year a three-year draft was imposed on all "physically fit" males. Upper Volta's quota of recruits in 1920 was larger than that of any other French West African colony.[54]

Taxation, forced labor, and conscription generated resistance and rebellion, especially in the west. Since the early days of colonization, the peoples of the west had displayed more resistance than the Mossi, as the colonial state was in sharper contrast to their lineage- and village-based societies than to the Mossi system. In 1897 the French destroyed the Bwa village of Ouarkoye between Bobo-Dioulasso and Dedougou when its inhabitants tried to disturb French supply lines.[55] There were also Samo revolts in 1898 and more clashes between the Bwa and the French in 1899. This is not to say that there were not also acts of rebellion among the Mossi. In 1899, after the death of their *naba* and the appointment of a new one, the Mossi of the Yatenga mounted a revolt that was quelled by the French. In 1908 a resistance movement started around Koudougou, where a Muslim leader asked Mossi not to pay taxes and organized a march onto Ouagadougou with 2,000 men armed with spears. After they were stopped some 40 km outside of town, the French retaliated by burning villages, seizing goods and animals, imprisoning some Mossi chiefs, dismissing others, and reducing the *mogho naba*'s colonial stipend by one-third. Skinner believes that this severe repression shocked the Mossi to the point that they "remained docile from that time onward, and thereafter the French were able to rule the country with one European administrator for every 60,000 Mossi."[56] Other groups, however, were repressed with equal violence. In spite of this, another Muslim-inspired rebellion occurred in May 1914 in the bend of the Black Volta. Its leaders were given long prison sentences in 1915.[57]

Although the heavy taxation had been the main reason for revolt up to that time, World War I brought about conscription and another bout of rebellion in the west, this time far more substantial.[58] In November 1915 the Marka village of Bouna announced that it would resist military recruitment. A few days later, after the Marka forced French recruiters to withdraw, the revolt quickly spread through the area, destroying the infrastructure and threatening colonial personnel. In late November the commander of the *cercle* of Bobo-Dioulasso and a local French-appointed chief were held under siege by Bwa and Marka insurgents for more than a week before they escaped and the rebels took over the town. In December the rebels attacked a Marka village loyal to the French, who later dislodged them. But on 23 December 1915 the Marka beat the French at the battle of Yankasso.

Stunned by the spread and the depth of the rebellion, the French began a campaign from Dedougou in February 1916 to crush the Marka, the Bwa, and other rebellious groups west of Bobo-Dioulasso. Meanwhile, in March, the revolts had spread north to Dori, where Tuareg insurgents attacked French troops. In May the French occupied Koudougou and in June defeated the Tuaregs. By the end of July, violent resistance to French rule had been put down.[59]

## The First Colony of Upper Volta and Its Partition

Following the revolts of 1915–1916, the French felt the need to reinforce colonial control. The region's backwater status also forced them to devise a plan to foster economic development: the creation of an incorporated colony, with its own governor, budget, and economic policies. The name *Upper Volta* was selected from among a pool that included *Volta-Niger, Nigerien Sudan, Volta,* and *Middle-Niger.*[60] The colony was made official on 20 May 1919, with Ouagadougou the capital of an area that included the *cercles* of Gaoua, Bobo-Dioulasso, Dedougou, Dori, Fada N'Gourma, Say, and the Cercle du Mossi. Upper Volta was to be a relatively insignificant colony. In 1920 it functioned with ten French administrators and fourteen indigenous adjuncts. Its first governor was Edouard Hesling, who was succeeded by Albéric Auguste Fournier in 1928. The Cercle du Mossi was split up little by little. In October 1920 Ouahigouya was made an independent *cercle*, in June 1921 it was Tenkodogo's turn, and in 1923 Ouagadougou became a *cercle* and the Cercle du Mossi was effectively dissolved. The rest of Haut-Sénégal-Niger became the Soudan (also called the French Sudan) in December 1920. In 1927 most of the *cercle* of Say was transferred to Niger.

The creation of Upper Volta did not spell relief for the Voltaics. Taxes continued to rise. In 1926 all males not in the military service became liable for public works. Labor and military recruitment continued in conditions that generated high mortality rates. Skinner cites a Mossi proverb that "white man's work eats [kills] people."[61] Governor Hesling often resorted to forced labor to support his plans to develop Upper Volta into a prosperous colony. His first action upon taking office was to have the *mogho naba* recruit 2,000 men to erect the colony's administrative buildings.[62] Infrastructure work to encourage the development of cotton demanded heavy labor, too. From 1919 to 1925 Hesling had 6,000 km of roads built. The colonial authority gave farmers free cotton seeds—and with them compulsory output figures. In 1924 a Service des Affaires Textiles was created. Around the same period the cultivation of rubber, earlier abandoned, was resumed, and groundnuts were introduced.

But Voltaic labor above all served the development of other colonies. In 1922 Upper Volta was asked to provide 6,000 workers for the Thies-Kayes railway and 2,000 men for the railways in Côte d'Ivoire.[63] Upper Volta thus acquired a reputation as a reservoir of labor. All in all, from about 1920 to 1930, 25,000 Voltaics worked on the Senegal rails and 55,000 on the Régie Abidjan Niger rails, mostly

*Workers on the Abidjan-Ouagadougou railway, October, 1954. Courtesy of the Centre National de la Recherche Scientifique et Technique.*

in Côte d'Ivoire. Agricultural laborers were sent to Sudan and Senegal, and others went to Côte d'Ivoire to work in the timber industry. A 1925 law providing for freedom of employment does not appear to have been seriously implemented in Upper Volta. Moreover, 45,000 men per year were recruited into the military in the 1920s so that conscription in France could be kept to a minimum.

The chiefs received a monthly salary and commissions on the collection of taxes. In 1934 an eighth-class district chief received Fr 600 per month, while a first-class chief made Fr 18,000 per month. Classes were a measure of seniority and merit; it took two to five years to move from one class to another upon nomination by the *cercle* administrator.[64] Thus chiefs became salaried workers of the French administration, with administrative and police functions. They continued to collect taxes and kept a list of the taxpaying members of the village. They registered births, deaths, marriages, and divorces; drew up census lists; implemented requisitions; and forwarded mail.[65] The *mogho* and Yatenga *nanamse* received an extra stipend. Village chiefs were given seeds, help in the construction of dams and wells, and medals and honors.[66] This policy crystallized the role of the *nanamse* and other chiefs as auxiliaries of the administration. But the "traditional" status of the chiefs was never reintroduced and continued to be diminished by policies such as Christian evangelization and improvement of the lot of women, which threatened their roles and functions (see Chapter 5).

In 1930 the worldwide economic depression took hold, and cotton, rubber, and groundnuts collapsed on the commodities market. With taxes, migration, poor rain, and locusts, the famine settled in by 1931. Upper Volta was dismantled on 5 September 1932 and split among the Soudan, Côte d'Ivoire, and Niger. The move followed an essentially economic rationality: favoring the migration of labor from the people-rich Mossi plateau to the resource-rich but people-starved coastal areas of Côte d'Ivoire and, secondarily, to the irrigated agricultural regions of the Niger basin in Soudan. Because of the economic failure of Upper Volta as a colony, its political assuagement, and its obvious comparative advantage in labor power, the French saw no reason to maintain it as a separate entity. Côte d'Ivoire received the lion's share of the former colony: the *cercles* of Tenkodogo, Gaoua, Batié, Ouagadougou, Bobo-Dioulasso, and part of Dedougou—populated by a total of 2 million people—to be called Upper Côte d'Ivoire. The other part of Dedougou and the *cercle* of Ouahigouya (712,000 people) joined Soudan, while Niger inherited the *cercles* of Fada N'Gourma and Dori (268,000 people).[67] With the Yatenga going to Soudan and the other kingdoms to Côte d'Ivoire, the Mossi found itself belonging to two different colonies, a situation its hierarchy would never accept. In 1938 Upper Côte d'Ivoire received some administrative autonomy and its own "resident superior," Edmond Jean Louveau, headquartered at Ouagadougou. (See Map 2.1.)

The first project for which Voltaic labor power was needed was the Office du Niger in Soudan, a rice- and cotton-growing scheme, which originally planned on receiving 1 million Voltaics but eventually only attracted a few hundred Mossi, Samo, and Marka. The needs of Côte d'Ivoire were more substantial. France sponsored the creation of Voltaic villages to attract seasonal migrants in 1933. They worked on cocoa and coffee plantations, in cotton fields, and on their own plots of food crops. The port of Abidjan and a railway meant to head north to Niger were built by Voltaics (the rail line reached Bobo-Dioulasso in 1934). The timber industry also heavily recruited Voltaics. Voltaic workers were officially volunteers but in reality had little choice. The colonial authorities of Côte d'Ivoire communicated their labor needs to the administrators of Upper Côte d'Ivoire, who then recruited the desired number of workers. It was not until 1936, with the coming of the Front Populaire in France, that the conditions of Voltaic migrants in Côte d'Ivoire improved.

When World War II broke out, the Mossi chiefs came out in favor of military recruitment. The *mogho naba* volunteered his two eldest sons, and more than 10,000 Mossi soldiers enlisted. Although usually referred to as the Tirailleurs Sénégalais, many of the French West African soldiers in the war were in fact Voltaic.[68] Yet numerous volunteers never left the colonies: Following the quick French defeat, Marshall Philippe Pétain capitulated and created the collaborationist Vichy Republic, while General Charles de Gaulle set up the French Free

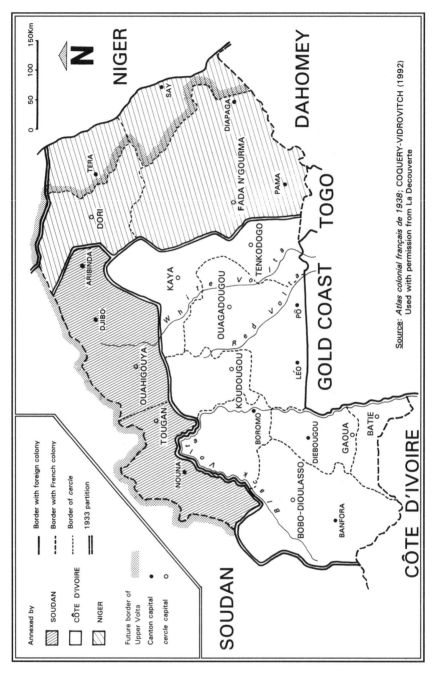

MAP 2.1  *The Partition of Upper Volta, 1933–1947*

"*E Pluribus Unum?*" *Precolonial and Colonial History* 27

Forces in London and called upon the colonies to join him against Germany and the Vichy government. Louveau chose to side with de Gaulle, invited the British to Ouagadougou, and helped people cross into Gold Coast to join the Free Forces. But the governor-general of French West Africa, Pierre François Boisson, declared his colonies' allegiance to Pétain; ordered Louveau to report to Dakar; and sent him to France to be tried for treason and sentenced to hard labor for life. Louveau escaped in 1943, however, and joined de Gaulle in Algiers. Boisson's decision to side with Vichy and to dismiss Louveau devastated the *mogho naba*. Adding insult to injury, his recruits ended up being used for forced labor. On 12 March 1942 Mogho Naba Kom II died, days after a total eclipse of the moon and hours after a meteor passed over Ouagadougou, apparently "vividly impress[ing] him."[69] The French concluded he died of heart failure, but rumors circulated that he committed suicide. His eldest son, Issoufou Congo, an officer in Côte d'Ivoire, became Mogho Naba Sagha II. On 8 November 1942 the Allies invaded North Africa. On 23 November Boisson proclaimed West Africa's allegiance to the provisional French government under General de Gaulle in Algiers.

## France's Fourth Republic and the Neutralization of Voltaic Radicalism

In early 1944 the Brazzaville conference of French colonial governors, championing the idea of progressive integration of the colonized within French society over that of emancipation, recommended to their government that colonies be granted a certain level of political representation in the metropolitan parliament after the forthcoming liberation of France. This recommendation was adopted in 1945, and sixty-one seats were created for overall "overseas" representation for the October elections for the Constituent Assembly of the Fourth Republic.[70] Although France's moral debt to the *tirailleurs* contributed in part to this modification of the colonies' status,[71] the representation of West Africa was limited to ten deputies (four of whom were Europeans), elected by two different constituencies, one for French citizens and the other for noncitizens. Upper Volta, which did not exist as a colony, did not have its own deputy. Yet the *baloum naba*, the (reportedly illiterate) palace intendant of the *mogho naba*, ran as a candidate from Côte d'Ivoire, with backing from the administration, against Félix Houphouët-Boigny.[72] The latter won with 13,750 votes (the *naba* garnering 12,900),[73] but when Houphouët-Boigny ran again for the new Constituent Assembly in June 1946, he was reelected unopposed. The constitution of the Fourth Republic having been approved in October, regular legislative elections were held in November to form a national assembly.[74] For these elections, voting rights were extended in French colonies to include wage earners, veterans, and property owners.[75] This time Côte d'Ivoire received three seats, and its Voltaic region made a massive entrance in metropolitan politics. Daniel Ouézzin Coulibaly (a Bobo-Fing) and Philippe Zinda

Kaboré (a Mossi opposed to the *chefferie,* the traditional Mossi hierarchy) were elected alongside Houphouët-Boigny under the colors of the Rassemblement Démocratique Africain (RDA), a transcolonial radical party created in Bamako, capital of Soudan, in October 1946. The RDA favored progressive emancipation and opposed colonialism and assimilation of the Africans by the French; it later allied with the Communist Party in the French parliament.[76]

The introduction of metropolitan politics in the colonies, together with the non-Mossi bias in the representation of Upper Volta, induced a Mossi current in favor of the re-creation of Upper Volta after World War II. The Mossi, having seen their empire partitioned in 1932, stood to gain more than other ethnic groups from a reunification of the colony. In 1945 several Mossi chiefs and some of their schooled subjects set up the Union pour la Défense des Intérêts de la Haute-Volta (soon to be transformed into the Union Voltaïque, or UV), to lobby for the reconstruction of Upper Volta. In July 1946 the *mogho naba* officially petitioned the French overseas minister to reestablish the colony. Although the minister replied favorably, it took another year to effect the change. On 4 September 1947 the French National Assembly passed a law reconstituting Upper Volta as a colony along the lines of its former territory, except for a northeast segment retained by Niger. In addition to appeasing the Mossi and serving certain French economic interests,[77] the reconstitution of Upper Volta supported the French policy of containing the spread of the RDA in West Africa and curbing Houphouët-Boigny's influence outside Côte d'Ivoire.[78] The years that followed would indeed be characterized by a struggle between Mossi-centered moderate parties allied to the French administration and the RDA, which—for a while—represented the federalist tendency among African political elites.

Albert Jean Mouragues, the first governor of reunified Upper Volta, entered office in April 1948 with the mission of furthering these goals.[79] Under pressure from Mouragues, Bishop Thévenoud warned his constituents of the incompatibility of Christianity and communism and supported the UV against the RDA in the June 1948 elections to the territorial and French national assemblies.[80] Amid allegations of electoral manipulation by the governor, UV's Henri Guissou (a Mossi), Mamadou Ouédraogo (a Mossi), and Nazi Boni (a Bwa-Bobo) were elected to the National Assembly,[81] this time giving the Mossi a representation in excess of their proportion in Upper Volta's population. The RDA failed to get a candidate elected but managed to send one representative to the Assembly of the French Union. A few months later the UV deputies contributed to the foundation of the parliamentary group Indépendants d'Outre-Mer (IOM), an anti-RDA, anti-communist group under the leadership of Senegal's Léopold Sedar Senghor.

With the re-creation of Upper Volta and the founding of the UV, the RDA's already limited influence among the Mossi continued to dwindle, but in the west, especially among the Bobo and the Lobi, it retained a fair level of popularity despite administrative harassment (Bobo-Dioulasso was chosen as the site of the second RDA congress in December 1948, but the meeting was forbidden by

*Mogho Naba Saaga II, August 1950. Courtesy of the Centre National de la Recherche Scientifique et Technique.*

Governor Mouragues).[82] In 1948 the Voltaic section of the RDA, called the Entente Voltaïque, still won eight seats at the elections to the general council, including that of first vice-president, and sent one delegate to the Assembly of the French Union in Paris and one to the grand council of the AOF. The UV won the majority. Nevertheless, in a shift from federalism to nationalism among West African political elites, the RDA was progressively marginalized at the Voltaic level as the UV rose in influence.

The 1950 breakup between Houphouët-Boigny and the French Communist Party and the RDA's parliamentary alliance with François Mitterrand's Union Démocratique et Socialiste de la Résistance (UDSR)—which represented a new shift, this time from radicalism to reform—did not prevent the colonial administration in Upper Volta from continuing its staunch opposition to the RDA, even though governors in other colonies had by then relaxed their attitude toward what had become a moderate party. The colonial manipulation of politics and long-term opposition to the RDA in Upper Volta may provide some clues to subsequent patterns of Voltaic politics, such as the blossoming of political parties and the failure to install a single party after independence, in contrast to neighboring countries, where the RDA provided the basis for single-party authoritarian regimes (e.g., Côte d'Ivoire, Guinea, Mali).

In the June 1951 elections, in which some women were for the first time allowed to participate, Guissou, Boni, and Mamadou Ouédraogo were reelected to the French national assembly, and Joseph Conombo, another Mossi and IOM member, won his first election. Again the RDA failed to send a deputy to Paris. Conombo, a *tirailleur sénégalais* in World War II, was representative of the new political aspirations of the Africans who fought in the war and came in contact with the French population and its political and cultural values.[83]

The marginalization of the RDA in Upper Volta became more apparent with the March 1952 elections to the Territorial Assembly. The UV won, with twenty seats; the Rassemblement du Peuple Français (RPF, an offspring of the UV led by a former French officer, Michel Dorange) was the runner-up, leaving the RDA without a single seat. In the October 1953 elections, the only RDA candidate was again defeated. Meanwhile, the UV members continued their ascent, and Conombo became Pierre Mendès-France's secretary of state for the interior in the French government in June 1954.

In addition to political change, the period following World War II witnessed considerable social evolution. Partly thanks to the influence of African legislators in Paris, several laws were passed that improved the status of the colonized. In December 1945 *indigénat* was abolished. In April 1946 a law ending forced labor, introduced by Houphouët-Boigny and bearing his name, was passed in parliament, and the French penal code was made applicable to sub-Saharan Africa. In May of the same year, a law sponsored by Lamine Gueye abolished the status of subject and colonized Africans became citizens—which did not, however, automatically grant them political rights. In September freedom of the press was recognized. In June 1950 equality in recruitment, pay, and promotion was established between Europeans and Africans overseas, and in December 1952 a new labor code for overseas France imposed a forty-hour week, paid holidays, and trade union rights.[84]

## Parties and Persons: Emergence of a Pattern

After Governor Mouragues left Upper Volta in 1953, the UV began to unravel. In 1955 it split between the Parti Social pour l'Emancipation des Masses Africaines (PSEMA), a de facto Mossi party founded by Conombo and Guissou and still favored by the colonial administration, and Nazi Boni's Mouvement Populaire d'Evolution Africaine (MPEA). The PSEMA had a large majority in the Territorial Assembly and controlled half the seats in the National Assembly. The rupture was both personal and regional, the MPEA representing the interests of the west and demanding a partition of Upper Volta into two territories, with increased autonomy for the region of the Bobo, Samo, Gurunsi, and Lobi. But the Mossi themselves began to lose their political cohesion. Younger and educated Mossi, under the leadership of Joseph Ouédraogo, former secretary of the Ouagadougou branch of the Confédération Française des Travailleurs Croyants (CFTC) union, were tired of the backwardness of "traditional" leaders and impatient for more independence.

The emergent political divisions among the Mossi coincided with a comeback of the RDA, which won the municipal elections in Bobo-Dioulasso in 1954. A pattern was established: The RDA enjoyed renewed popularity in the western part of the country, while Mossi-dominated parties remained strong in the center and the east. This RDA did not have much in common with its pre-1950 ancestor; it had become more an instrument of Houphouët-Boigny's particularist policies than the agent of West African unity.

Conombo and Guissou were reelected for the PSEMA in the January 1956 French legislative elections, Nazi Boni won his seat back for the MPEA, and Gérard Kango Ouédraogo won the fourth seat. Gérard Kango's party was the Mouvement Démocratique Voltaïque (MDV), an offspring of the RPF, and it was coheaded by Dorange. Gérard Kango and Dorange both represented Ouahigouya, the capital of the Yatenga kingdom, and reflected a further division among the Mossi, based on kingdoms this time. The RDA, whose Voltaic branch was now the Parti Démocratique Voltaïque (PDV) came out relatively strong in the west but fell short of winning a representative in the National Assembly, nevertheless getting nine seats in the Territorial Assembly.

In 1956 the *loi cadre* allowed for universal suffrage in a single electoral college and gave the territorial assemblies ministerial powers over each colony. The federations of AOF and AEF thus became irrelevant political entities, and the idea of federalism in West Africa definitely took a backseat.[85] In this new context the PDV and the PSEMA united to form the Parti Démocratique Unifié (PDU) in November 1956, with Conombo as president. Ouézzin Coulibaly, making a comeback to Voltaic politics following the balkanization of Africa consecrated by the *loi cadre* and the reformation of the RDA, and sent by Houphouët-Boigny to reorganize the Voltaic branch of the RDA, became vice-president, although he was still a deputy from Côte d'Ivoire in the French National Assembly. With Nazi Boni's MPEA weakened in the west by the resurgence of the RDA, the principal opposition to the PDU became Gérard Kango Ouédraogo's MDV.

In the March 1957 elections to the Territorial Assembly, the first ones since the *loi cadre*, the PDU won thirty-seven seats,[86] the MDV polled twenty-six, the MPA five, and independent candidates two. The strong showing of the PDU in the west and its defeat by the MDV in other parts of the country were a setback for the Mossi-based PSEMA of Conombo. The PDV component of the PDU had won thirty-two seats compared to only five for the PSEMA. It was thus Ouézzin Coulibaly who was elected vice-president of an executive council[87] that counted seven PDU and five MDV members. The portfolio of agricultural economy was held by Maurice Yaméogo, a postal clerk and CFTC unionist.

The March 1957 elections were followed by a series of successive political maneuvers based on individual alliances and arithmetics rather than political choices. These would have long-lasting consequences and were in fact the first indication of the future pattern of parliamentary party politics in Upper Volta and Burkina Faso. Shortly after the formation of the Coulibaly "government," Conombo and his PSEMA, no longer content to play second fiddle to the PDV and uneasy with Coulibaly's desire to turn the PDU into a territorial section of the RDA, seceded from the PDU to form the Groupe Solidarité Voltaïque (GSV) with their former opponents, the MDV and the MPA—thereby forcing a change of majority in the territorial assembly and securing a vote of defiance against Ouézzin Coulibaly. The latter refused to resign, however, and obtained instead the defection of four GSV members by offering them a ministerial portfolio, provoking a new majority shift, this time in favor of his own PDV. Among the defectors was Maurice Yaméogo, appointed minister of the interior in the new, homogeneous PDV government. With the PDV on the one hand and the PSEMA, MDV, and MPA on the other, the earlier divisions between RDA and Mossi-based parties had resurfaced. Interestingly enough, the defector Yaméogo was a Mossi from Koudougou and as such was reputedly opposed to the Ouagadougou *chefferie*.

In March 1958 the GSV joined the new umbrella group Parti du Regroupement Africain (PRA), and the country slid toward a bipartisan system. Yet in Virginia Thompson's words, "The term 'party' could not be properly applied to the cliques that formed and reformed around outstanding personalities."[88]

## Reluctantly Moving Toward Independence

The September 1958 referendum on the French Fifth Republic that Charles de Gaulle used to reinforce his presidential powers was, for West African colonies, a referendum on whether to remain as republics within the French Community—the new name de Gaulle had given to the association of France and its colonies—or to become independent countries. The French Community gave its "member republics" autonomy and self-government, but the organs of the community, that is, the French government, retained control over foreign policy, defense, and economic matters, leaving only social affairs in the hands of the republics. Membership in the community was a sovereign decision: Guinea voted against it and immediately became independent in 1958, though with much French resentment and amid many French-made obstacles. Republics could request a "transfer of competence" from the community to themselves at any time. This was how they acquired independence from April to October 1960. Logically, the adoption of the community implied the end of African representation in the French national assembly. It also transformed the Assembly of the French Union into the Senate of the Community, which had no more real powers than its predecessor and hardly ever met. The colonies' territorial assemblies were to become territorial legislative assemblies, the vice-presidents of the councils of ministers becoming presidents, while governors would henceforth be called high commissioners representing the president of the community (i.e., de Gaulle). In Upper Volta, where both the PRA and RDA campaigned in favor of the community, the referendum was approved on 28 September by 98.9 percent of the voters.

A few days earlier, on 7 September 1958, Ouézzin Coulibaly had died in Paris, where he had been in treatment since July. His interior minister, Maurice Yaméogo, succeeded him as acting vice-president of the executive council. With Coulibaly and Zinda Kaboré dead and Houphouët-Boigny in Côte d'Ivoire, none of the first-generation politicians of Upper Volta remained at the inception of independence. Unlike other colonies, such as Côte d'Ivoire, Senegal, or Mali, where the first actors of postwar politics were still at the forefront, Upper Volta's leaders belonged to the second generation of politicians. Their relative inexperience and lack of social rooting contributed to the country's dawning instability and may have prevented Yaméogo from successfully imposing single-party politics in Upper Volta. From this perspective, the consequences of the death of Coulibaly, the first politician to have—if only temporarily—united Mossi and non-Mossi within a single party, may have been far-reaching indeed.

The *mogho naba* perceived Upper Volta's change of status as an appropriate time to push for his idea of constitutional change. Mogho Naba Saaga II had died in 1957, and the new *naba* was his twenty-seven-year-old son, Mogho Naba Kougri (literally, "the Rock"). The latter saw his power wane with the rise of the new Mossi elites and the dissolution of Mossi identity in the nascent Voltaic one. He dreamt of a constitutional monarchy of which he would be the king. On 17 October 1958, two days after having unsuccessfully requested politicians to form a government of national unity and apparently with the support of Conombo and his PSEMA,[89] Mogho Naba Kougri sent 3,000 warriors "armed with arrows and machetes"[90] in a youthful attempt to surround the territorial assembly and force it to comply with his political project. They quickly retreated, however, upon the intervention of French troops.[91] High Commissioner Max Berthet regretted this "unacceptable and incomprehensible" demonstration, and the government council issued a communiqué the next day "stigmatizing" the *mogho naba*'s action.[92]

On 20 October Yaméogo was elected president of the Council of Ministers in his own right by the Territorial Assembly. On 9 December he formed an RDA-RPA cabinet of national unity, thereby co-opting several opposition leaders and essentially creating an opposition vacuum. Two days later Upper Volta officially became an autonomous republic within the French Community, with Yaméogo as president of its Council of Ministers. Its new constitution was adopted on 28 February 1959 and approved on 15 March 1959.

The main political debate in French West Africa in the years 1958–1960 centered on the relations that the newly autonomous republics should enter with one another and with France. This was but a continuation of the debates that began in 1946 between proponents of federalism and nationalism and between supporters of emancipation and those willing to work within the French colonial structures in the spirit of integration. By the late 1950s Senghor of Senegal and

Modibo Keita of Soudan represented the federalist trend and assumed the mantle of emancipation. Houphouët-Boigny was the nationalists' and integrationists' flag carrier.

The first debate, between federalism and nationalism, revolved around two different types of association among the colonies: the Fédération du Mali, which envisioned the federation of colonies into a West African state, and the Conseil de l'Entente, which foresaw simple cooperation among independent states. Still insecure in his new powers and essentially deprived of both vision and ideology, Yaméogo wavered according to his changing perception of his interests. In December 1958, while Modibo Keita visited Ouagadougou to lobby in favor of the federation, Yaméogo was reported to have told de Gaulle in Paris that he favored "closed cooperation with [Côte d'Ivoire] over unconditional membership in a West African federation."[93] Yet in January 1959 Yaméogo was heading the Voltaic delegation at the Dakar Federal Constituent Assembly that proclaimed the Federation of Mali, and on 28 January the Voltaic territorial constituent assembly ratified the constitution of the Mali Federation. A month later the constitution of the Republic of Upper Volta was approved without any mention of membership in the Mali Federation. In the end only opposition politicians from the west, such as Nazi Boni, were present at the inauguration of the federation in Dakar on 24 March 1959. In April, days after signing a protocol of economic cooperation with Côte d'Ivoire, Yaméogo joined the Conseil de l'Entente with Côte d'Ivoire and Niger. Yaméogo's decision marked the first strong manifestation of Houphouët-Boigny's involvement in and influence on politics in Upper Volta and institutionalized Upper Volta's dependence on Côte d'Ivoire.

At the April 1959 legislative elections, the PDV won sixty-nine seats out of seventy-five, leaving the remaining six seats to the PRA. The PDV was now well entrenched in the Mossi community (Yaméogo had forced the *mogho naba* to drop his support for the Mali Federation), but the party nevertheless resorted to electoral manipulation to enhance its chances of success. Indeed, an ordinance taken by Yaméogo on 9 March 1959 provided for a one-round majority poll in districts of fewer than 300,000 inhabitants and for proportional representation in those of more than 300,000 inhabitants. According to Philippe Lippens, "this system applied majority elections were the RDA dominated and proportional where it was a minority, [providing the RDA] with a clear advantage."[94]

Strengthened by his new majority, Yaméogo was reelected president of the Council of Ministers and rapidly imposed authoritarian rule. Having obtained absolute powers,[95] he dismissed the mayor of Ouagadougou, the PDU's Joseph Ouédraogo, in August, allegedly for "malfeasance in office"[96] but more probably because he had become a political foe. In October opponents of Yaméogo and supporters of the Mali Federation formed the Parti National Voltaïque (PNV) under the leadership of Nazi Boni. Two days later Yaméogo ordered the PNV dissolved. Nazi Boni then created the Parti Républicain pour la Liberté (PRL). But Boni was about to learn that the nature of politics in Upper Volta had changed under Yaméogo, and in January 1960, days after Yaméogo became president of the Republic of Upper Volta, he outlawed the PRL. In June six opposition figures, including Nazi Boni, sent a letter to the president "suggesting that all of the people should be invited to help build the new country."[97] Five of them were arrested a few days later. Nazi Boni escaped to Mali; he lived in exile there and in Senegal until 1966.[98] The other five were released in November 1960.

With the issues of Yaméogo's power and the Conseil de l'Entente settled, attention shifted to the second debate, between emancipation, as expressed by independence, and continued integration with France. Because the leaders of the West African nations who did not believe in federalism (Côte d'Ivoire, Dahomey, Niger, and Upper Volta), had little to gain from independence, they did not eagerly await it. François Bassolet quotes Yaméogo declaring to High Commissioner Berthet in 1959: "There are lunatics who dare ask for independence. We cannot even build matchboxes and they want us to be independent. . . . We at the RDA, we do not care for independence."[99] But as Patrick Manning has pointed out, "the argument for independence came more from outside francophone sub-Saharan Africa than from within."[100] With the independence of Indochina, the Algerian civil war, the independence of Ghana, and the Guinean no to France, "the logic of independent African nations had become inexorable."[101] Yet senior West African political leaders remained committed to the end, though in different ways, to the idea of a community between France and Africa. Their concern was to increase their sovereignty and their equality within this community, not to break away from it. The French concepts of assimilation and association had successfully permeated the minds of the colonized.[102] African deputies, for the most part, truly believed in the notion of an eventual multinational French Union in which Africans would be equal French citizens. As Michael Crowder colorfully puts it: "Right up until 1956 most of the activities of the African deputies in the National Assembly were concerned not with the achievement of local self-government but with squeezing as much juice as possible out of the assimilationist lemon."[103]

After the introduction of the *loi cadre* in 1956, the IOM and the RDA pushed for federation with France. For Senghor and the IOM, the federated units had to be France, the AOF, and the AEF, not the specific colonial territories. Houphouët-Boigny wanted to bypass the AOF and federate France with the individual territories because he did not want Côte d'Ivoire to subsidize an AOF government in Dakar and because he feared the radicalism of non-RDA politicians.[104]

It was a younger generation of politicians, students in France in the late 1950s, who brought the issue of independence to the forefront while older politicians remained absorbed by the question of the form of federation with France. Later, when the question became more pressing, the issue of the economic viability of their states became the politicians' main concern, since de Gaulle had made it clear that independence would involve the withdrawal of French assistance. West African colonies were indeed receiving large amounts of French aid, their agricultural production was substantially subsidized, and they generally lived above the means they otherwise could have afforded.

Mali and Senegal, embittered by the failure of the Federation of Mali (which they alone joined—and which would soon collapse), were the first to begin negotiating for independence. Seeing that de Gaulle would no longer oppose Mali and Senegal's wish, Houphouët-Boigny followed their example, with Upper Volta in his footsteps. In the perspective of politics since 1946, independence had singularly little relevance in the political evolution of Upper Volta. By the time it was set free from France, the main patterns of Voltaic politics were already well entrenched, and Yaméogo's power was as solid as it would ever be. Nor did independence reduce the influence of France, either politically or economically. For a country such as Upper Volta, it can reasonably be argued that the reforms of the Brazzaville conference and the *loi cadre* had further-reaching consequences both for the political elites and the population at large than did independence, which was in many ways merely a juridical change.

## Notes

1. There are almost as many spellings of *Mossi* as there are authors writing on them. Jean Audouin and Raymond Deniel spell the term for the people *Mossi* and for their leader *moog-naaba,* stating that it is the "currently used orthography in vulgarization books"; Jean Audouin and Raymond Deniel, *L'Islam en Haute-Volta à l'époque coloniale* (Paris: L'Harmattan and INADES, 1978), 5. According to A. A. Dim Delobson (*L'Empire du Mogho Naba: Coutumes des Mossi de Haute-Volta* [Paris: Donat-Montchrestien, 1932]), the term *Mossi* is a French perversion of *Mossé,* the plural of *Moaga,* the inhabitant of the *Mogho* (the land of the Mossé). The king is the *mogho naba.* Claudette Savonnet-Guyot, "Le Prince et le Naaba," *Politique Africaine,* 20 (December 1985): 30, and Michel Izard, *Gens du pouvoir, gens de la terre* (Cambridge: Cambridge University Press, 1985) choose *Moose* and mention decree 75/PRES/EN of 16 December 1975, which sets the official orthography as *Mooga* (singular), *Moose* (plural), and *Moogo* (the territorial entity formed by the Moose kingdoms). The problem with Izard and Savonnet-Guyot's orthography is that it gives little indication of the pronunciation of the words. For the sake of simplicity and consistency, I follow the orthography used by the first authority on the subject in the United States, Elliott Skinner of Columbia University, who uses *Mossi* for both singular and plural and *mogho naba* for their emperor. But I wish to warn the reader of the many variants in spelling Burkinabè names in general.

2. Research on the Mossi owes much to Maurice Delafosse, Michel Izard, A. A. Dim Delobson, and Elliott Skinner. I hope to offer here a synthesis of their findings.

3. That the Mossi system was disrupted nearly as soon as it was discovered is in part responsible for this lack of a definitive history. See Dominique Zahan, "The Mossi Kingdoms," in Daryll Forde and P. M. Kaberry, eds., *West African Kingdoms in the Nineteenth Century* (London: Oxford University Press for the International African Institute, 1967), 163.

4. Delafosse and Dim Delobson place the kingdoms' beginning in the eleventh century; see Maurice Delafosse, *Le Pays, les peuples, les langues, l'histoire, les civilisations du Haut-*

*Sénégal-Niger,* 3 vols. (Paris: Larose, 1923) and Dim Delobson, *L'Empire.* Izard and Skinner date it to the fifteenth century; see Izard, *Gens,* and Elliott P. Skinner, *The Mossi of Burkina Faso: Chiefs, Politicians and Soldiers* (Prospect Heights, Ill.: Waveland Press, 1989). Delafosse and Izard deal primarily with the Yatenga kingdom, while Dim Delobson and Skinner study the Ouagadougou kingdom.

5. One possible explanation for the wide discrepancy in dates is that two different groups have been referred to as the Mossi. According to Daniel M. McFarland, there are the northern Mossi, who, "after crossing the Niger between modern Niamey and Say, moved north and west via modern Dori and Aribinda between the thirteenth and fifteenth centuries. These are the Mossi who most likely came into contact with Mali and Songhai" (Daniel Miles McFarland, *Historical Dictionary of Upper Volta* [Metuchen, N.J.: Scarecrow Press, 1978], 112) and could actually be the ethnic groups preceding the foundation of the kingdom of Yatenga. The southern Mossi are those who originated from Ghana in the fifteenth century, as this chapter shows. For lack of written historical documents, a fair amount of imprecision must be accepted.

6. Dim Delobson, *L'Empire,* 1–11; Albert S. Balima, *Genèse de la Haute-Volta* (Ouagadougou: Presses Africaines, 1970); and Skinner, *The Mossi,* 7–8 and 239–240.

7. Dim Delobson, *L'Empire,* 5.

8. A world defined as the area between the White and the Black Voltas.

9. Dim Delobson, *L'Empire,* 11.

10. Ibid., 1.

11. Skinner, *The Mossi,* 9. Yet both Skinner and Dim Delobson, who seem to think that the Ouagadougou kingdom was founded in the late tenth century, stress that it is only with the fifteenth ruler, Mogho Naba Sana (ca. 1430–1450) that Ouagadougou became "the capital of the dynasty." This thus brings them in much closer concordance with the 1495 date Izard suggests for the foundation of Ouagadougou.

12. This rivalry is best illustrated by the fact that the Mossi from Yatenga called themselves Mossi and called the Mossi from Ouagadougou Gurunsi, while the Mossi from Ouagadougou called themselves Mossi and labeled the inhabitants of Yatenga Yadse. Izard, *Gens du pouvoir,* 2.

13. Skinner, *The Mossi,* 10.

14. Ibid., 9. See also Larba Yarga, "Modernisation administrative et autorité traditionelle en Haute-Volta" (Université de Nice, 1975, mimeographed), 22.

15. Izard, *Gens du pouvoir.*

16. Skinner, *The Mossi,* 15; Skinner spells it *nam.*

17. Izard, *Gens du pouvoir,* 18.

18. It also shows the conquerors' understanding that they needed to preserve and assimilate the agricultural producers if they were not to go hungry.

19. Savonnet-Guyot, "Le Prince," 31.

20. Izard, *Gens du pouvoir,* 311, and Claudette Savonnet-Guyot, *Etat et sociétés au Burkina: Essai sur le politique africain* (Paris: Karthala, 1986), 115. See also Zahan, who stresses how "[the] nakombse and the state officials [were] constantly recruiting new families within their sphere of influence. These [were] given Mossi patronymics and thus [became] Mossi, soon to lose former family and religious ties"; "The Mossi Kingdoms," 156.

21. Plural of *naba.*

22. Zahan, "The Mossi Kingdoms," 168–170.

23. Izard, *Gens du pouvoir,* 29–30 (for the village level) and 411 (for the Yatenga *naba*).

24. Ibid., 407.

25. Savonnet-Guyot, *Etat,* 111. Zahan stressed the point before Savonnet-Guyot: "[The] degree of autonomy and subordination of the Mossi principalities vis-à-vis the great Ouahigouya and Ouagadougou kings entailed a relationship which was quite distinct from that between feudal vassal and suzerain"; "The Mossi Kingdoms," 154.

26. A mistake repeatedly made by some Burkinabè leaders between 1983 and 1991 in their eagerness to apply Marxist-inspired historical analysis to their national situation.

27. See Dim Delobson, *L'Empire,* 86–96.

28. Skinner, *The Mossi,* 116.

29. Ibid., 127.

30. Ibid., 122.

31. As paraphrased by Savonnet-Guyot, *Etat,* 123.

32. Louis Binger, *Du Niger au Golfe de Guinée,* vol. 1 (1892), 459–468 and 498–502, quoted in John D. Hargreaves ed., *France and West Africa: An Anthology of Historical Documents* (London: Macmillan, 1969), 165–171.

33. Skinner, *The Mossi,* 127.

34. Note, however, that Dim Delobson was an agent of the French colonial administration and therefore was probably biased in favor of French "pacification."

35. This is not to say they had no chiefs. They most often did. But they had no institutionalized chiefdoms, no separation of polity and society.

36. Savonnet-Guyot, *Etat,* 27–39.

37. On the Samo, see Françoise Héritier, "La Paix et la pluie, rapports d'autorité et rapport au sacré chez les Samo," *L'Homme,* 13, 3 (1973): 123–135. On the Senufo, consult Bohumil Holas, *Les Sénoufo* (Paris: Presses Universitaires de France, 1957). As for the Bwa, I follow again the presentation of Savonnet-Guyot, *Etat,* explicitly derived from the work of Jean Capron, *Communautés villageoises Bwa: Mali-Haute-Volta* (Paris: Institut d'Ethnologie, 1973).

38. Maurice Duval, *Un totalitarisme sans Etat: Essai d'anthropologie politique à partir d'un village burkinabè* (Paris: L'Harmattan, 1985), 28.

39. Skinner, *The Mossi,* 5.

40. Savonnet-Guyot, *Etat,* 125.

41. On marriage especially, see Elliott P. Skinner, "Processus de l'incorporation politique dans les sociétés africaines traditionelles: Le cas des Mossi," *Notes et Documentations Voltaïques,* 1, 4 (July-September 1968): 29–47.

42. Chapter 5 addresses the possible contemporary extension of Mossi assimilationism to the west and questions whether at a time of relative economic hardship it is likely to breed unity or antagonism.

43. Catherine Coquery-Vidrovitch, ed., *L'Afrique occidentale au temps des Français. Colonisateurs et colonisés, 1860–1960* (Paris: La Découverte, 1993), 253.

44. Dim Delobson, *L'Empire,* 32–33.

45. Coquery-Vidrovitch, *L'Afrique occidentale,* 255.

46. Dim Delobson, *L'Empire,* 40. Dim Delobson recounts that the British had given the *mogho naba* a British flag as a signal of the British claims but that it could not be found when the Voulet-Chanoine mission approached because it had been used to make clothes. Ibid., 38.

47. Skinner, *The Mossi*, 157.
48. R. L. Buell, *The Native Problem in Africa* (New York, 1928), 1019, as reproduced in Hargreaves, *France and West Africa*, 223.
49. Yarga, "Modernisation," 14.
50. Ibid., 20.
51. Coquery-Vidrovitch, *L'Afrique occidentale*, 261.
52. Ibid., 266.
53. Ibid., 266–267.
54. McFarland, *Historical Dictionary*, 24.
55. Ibid., 120.
56. Skinner, *The Mossi*, 158.
57. McFarland, *Historical Dictionary*, 21–22.
58. For an account of the 1915–1916 revolts, see Lazoumou Seni, *La Lutte du Burkina contre la colonisation: Le cas de la région ouest, 1915–1916* (Ouagadougou: Imprimerie des Forces Armées Nationales, 1985).
59. Coquery-Vidrovitch, *L'Afrique occidentale*, 268–269.
60. Ibid., 270.
61. Skinner, *The Mossi*, 160.
62. Ibid., 163.
63. Ibid.
64. Yarga, "Modernisation," 48.
65. Ibid., 45.
66. Skinner, *The Mossi*, 165.
67. Ibid. 175, for population figures.
68. For an account of a Burkinabè's experience as a *tirailleur* in World War II, see Joseph Issoufou Conombo, *Souvenirs de guerre d'un "Tirailleur Sénégalais"* (Paris: L'Harmattan, 1989).
69. As quoted in a French report cited by Skinner, *The Mossi*, 179.
70. François Borella, *L'Évolution politique et juridique de l'Union Française depuis 1946* (Paris: Librairie Générale de Droit et de Jurisprudence, 1958), 37.
71. Indeed, according to Michael Crowder, "at one stage in the war, more than half of the Free French troops were of African origin." Michael Crowder, *Colonial West Africa: Collected Essays* (London: Frank Cass, 1978), 284–285.
72. Houphouët-Boigny, who represented the interests of African planters against their French counterparts, was perceived as a threat by the French administration, which was intent on preventing his election.
73. According to McFarland, *Historical Dictionary*, 32. According to Ruth Morgenthau, who quotes *Le Monde*, Houphouët won on the second ballot with 12,980 votes to the *baloum naba*'s 11,621. Morgenthau also quotes a description by Houphouët of the *baloum naba*'s telling his supporters, "Vote for my friend Houphouët.... What will I do in that Paris, with my forty-four wives and at my age? I can't speak a blasted word of French!" Ruth Schachter Morgenthau, *Political Parties in French-Speaking West Africa* (Oxford: Clarendon Press, 1964), 179.
74. The constitution of the Fourth Republic created three metropolitan assemblies with African participation: the National Assembly of Deputies (which had from 586 to 627 members, four of whom represented Upper Volta), which was the effective center of power;

the Council of the Republic, an indirectly elected senate with four Voltaic senators; and the Assembly of the French Union, a consultative body composed half of French metropolitan citizens and half of overseas "deputies" with eight Voltaic members. In the colonies the constitution set up general councils or territorial assemblies, elected by two electoral colleges (one European and one African) until 1956, when a single electoral college was adopted. Upper Volta's territorial assemblies had fifty members, ten elected by the first electoral college and forty by the second. Territorial assemblies were elected in 1948, 1952, and 1957. Last and least, there was a grand council of the AOF, based in Dakar, where each colony sent five representatives.

75. McFarland, *Historical Dictionary,* 33.

76. Africans elected to the French parliament sat with different political groupings: socialists, communists, or republicans. While the idea of a large West African party originated among deputies from different groupings, the socialists and the republicans were discouraged from further exploring this option by their French mentors under pressure from the government. Thus only communist deputies were present at the Bamako conference, the most notable of whom was Houphouët-Boigny.

77. See the section entitled "The Colonial Economy" in Chapter 4.

78. The death of Philippe Zinda Kaboré under mysterious circumstances in Abidjan in May 1947, shortly after he had delivered a virulent speech against the Mossi *chefferie,* contributed to this French goal. Indeed, the French administration was eager to bar the RDA from keeping Kaboré's seat and concluded that if Upper Volta could be reunified before the partial elections for this seat, it could be won by a candidate the *mogho naba* appointed. The *mogho naba* was thus encouraged to sever ties with the RDA against the promise of the reunification of Upper Volta. Indeed, although the French had taken no action since the first requests for reunification, the death of Kaboré was followed in June by a proposition of law for the reunification of Upper Volta introduced by MPR senators Alain Poher and Henri Guissou, approved in July by the Council of Ministers, adopted without debates by the assembly in August, voted upon on 4 September and published in the *Journal Officiel* on 5 September. See Coquery-Vidrovitch, *L'Afrique occidentale,* 280–281.

79. Later governors were Louis Eugène Geay (acting, March-October 1950), Salvador Jean Etcheber (1953–1956), Yvon Bourges (acting, 1956–1958), Max Berthet (acting, 1958), and Paul Masson (high commissioner, 1958–1960).

80. The first elections to territorial assemblies were held in other colonies in December 1946 for a six-year term. As Upper Volta did not then exist, its first elections were held in 1948 for a four-year term.

81. Ouézzin Coulibaly was temporarily continuing his political career in Côte d'Ivoire.

82. Virginia Thompson and Richard Adloff, *French West Africa* (Stanford, Calif.: Stanford University Press, 1957), 175.

83. He later married a French woman.

84. These measures were introduced for all sub-Saharan French colonies.

85. Patrick Manning, *Francophone Sub-Saharan Africa, 1880–1985* (Cambridge: Cambridge University Press, 1988), 147–148.

86. But Guissou was defeated by Maurice Yaméogo, an independent candidate, in the Koudougou area.

87. The governor of the colony was de jure president until 1958.

88. Virginia Thompson, *West Africa's Council of the Entente* (Ithaca, N.Y.: Cornell University Press, 1972), 8.

89. François D. Bassolet, *Evolution de la Haute-Volta* (Ouagadougou: Imprimerie Nationale, 1968), 78–79.

90. Ibid.

91. For further developments and analysis of this episode, see "Ethnicity and Language," Chapter 5.

92. As quoted in Skinner, *The Mossi*, 202.

93. McFarland, *Historical Dictionary*, 40.

94. Philippe Lippens, *La République de Haute-Volta* (Paris: Berger-Levrault, 1972), 19.

95. He had already received full powers for three months from the Constituent Assembly in January 1959.

96. McFarland, *Historical Dictionary*, 125.

97. Ibid., 42.

98. Nazi Boni eventually returned to Ouagadougou in 1966 after Yaméogo's overthrow and died in a car accident in 1969.

99. Bassolet, *Evolution*, 73.

100. Manning, *Francophone*, 147.

101. Ibid., 149.

102. The French themselves never seemed to believe in their own policy of assimilation for fear that France might become "a colony of her own colonies," in the words of National Assembly speaker Edouard Herriot, as quoted in Crowder, *Colonial West Africa*, 285.

103. Michael Crowder, "Independence as a Goal in French West African Politics: 1944–1960," in ibid., 287.

104. Ibid., 298.

# 3

# POLITICAL INSTABILITY AND THE QUEST FOR LEGITIMACY

The instability of the Burkinabè state is nowhere more visible than in its politics. On five occasions since 1966, the military or factions thereof have seized power either from civilians or from other soldiers, and there has never been a peaceful transition from one competitively elected regime to another. The political system has shifted back and forth from multiparty to single party to no party under four different constitutions or no constitution at all. Political parties have appeared, merged, and vanished. Politicians have risen to prominence only to be overthrown, imprisoned and later rehabilitated, or murdered. Trade unions have opposed virtually every regime and have been battered by almost as many. In a nutshell, there has never been any social consensus on the state, much less on the form of government. Regimes have endlessly sought a legitimacy that has remained elusive beyond the short term.

This chapter reviews the symptoms of this absence of legitimacy in Burkina's political evolution since independence, further identifies the characteristics of its polity, and suggests roots to its political unsteadiness. (Additional variables accounting for its overall instability are left, however, for Chapter 5.)

## The First Republic

Upper Volta became independent within weeks of many other French colonies, on 5 August 1960. It started its life as a sovereign country with a multiparty system and a regime inspired by the French liberal democratic model. But much like other former French colonies, its system was soon corrupted.

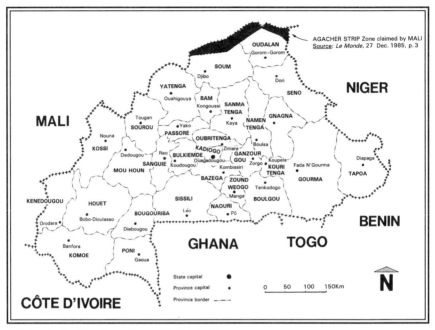

MAP 3.1   Burkina Faso: General Political Map

## The Drive Toward Authoritarianism[1]

Upper Volta had been a de facto bipartisan system since 1958, with the PRA the only remaining sizable opposition to the RDA. But with the participation of the main PRA leaders in successive Yaméogo governments, their party soon faltered as a credible alternative to the president. It was thus with a government in total control of the legislative power that Upper Volta became independent. On 27 November 1960, 99.5 percent of voters approved a new constitution that reinforced presidential powers, set up a unicameral parliament, but still recognized multiparty politics. On 8 December Yaméogo was elected president under the new constitution. At municipal elections in the same month, the president formally imposed the one-party system ("The only party that legislates in Upper Volta is the RDA. Everybody knows it."), preventing other parties from competing, so as "not to project abroad the sad and demagogic impression that we are divided at home."[2] Yaméogo took this step without a revision of the constitution, and several politicians who opposed it were arrested. Most of them were later released, and many joined the ranks of the president's supporters. It was thus within months of independence that the RDA had become de facto the country's single party, a strikingly rapid evolution away from the intentions of the constitution.

As political opposition was eliminated, so, too, was the institutional pluralism of the state, a situation that benefited the executive power and the RDA leadership. In December 1960 the speaker of the National Assembly, Begnon Koné, described his vision of the parliament's role: "to allow the executive power to take, in complete calm and serenity, the measures needed for Upper Volta's progress, prosperity, and grandeur."³ At the February 1962 RDA congress, the party declared its supremacy over all national institutions. The next step was to change the RDA into a mass organization, its branches reaching all the way down to the village committees under the organizational principle of democratic centralism, in which the upper units control the lower ones. The last stage of this evolution was the concentration of power into the hands of Yaméogo, both as head of the RDA and as president of the country.

A Mossi from Koudougou, Maurice Yaméogo was born in 1921 and went to the secondary school (*petit séminaire*) run by the Catholic mission at Pabré. He joined the colonial administration and was unionized with the CFTC. He became the Koudougou representative at the Territorial Assembly in 1946 as a member of the UV and remained a member in each territorial assembly until Ouézzin Coulibaly asked him to join his government in 1958. Although an opportunistic politician, Yaméogo did not believe in democracy and lacked common sense and restraint on many occasions. On national radio in 1965 he called Guinean president Ahmed Sékou Touré a "bastard among bastards" and "the prototype of the most intolerable immorality."⁴ According to McFarland, a usually moderate commentator of Upper Volta's history, Yaméogo "almost eliminated his opposition" while in power, came "close to bankrupting the nation with his presidential perquisites," and "cut himself off from reality by surrounding himself with sycophants who told him only what he wanted to hear."⁵

## Trade Unions and the Fall of Yaméogo

Yet Yaméogo's authoritarian tendencies failed to reach totalitarian proportions, in part because of the success of the trade unions in remaining independent from the regime.⁶ With political parties other than the RDA all but banned, opponents resorted to trade unions to voice their concerns. At the time of its independence, Upper Volta counted three major federations of trade unions: the Union Syndicale des Travailleurs Voltaïques (USTV), affiliated to the radical Union Générale des Travailleurs d'Afrique Noire (UGTAN) present in other former French colonies; the moderate Confédération Africaine des Travailleurs Croyants (CATC); and the Union Nationale des Syndicats des Travailleurs de Haute-Volta (UNSTHV), a member of the reform-minded Confédération Internationale des Syndicats Libres (CISL). Joseph ki-Zerbo, a staunch opponent of Yaméogo who in Dakar in 1958 had founded a socialist-inspired party, the Mouvement de Libération Nationale (MLN), had since joined the USTV.⁷ Joseph Ouédraogo, the former mayor of Ouagadougou, was active in the CATC.

Unable to co-opt the trade unions, Yaméogo undertook to curb their power, and on 24 April 1964 the National Assembly outlawed strikes and limited workers' rights. This proved insufficient for his purpose, however, as it was a confrontation with unions that brought about the relatively sudden downfall of his regime. The sequence of events highlights the superficiality of Yaméogo's power and the irrelevance of electoral success and parliamentary control to actual power and legitimacy.

On 3 October 1965, 99.97 percent of the voters reelected Yaméogo president. On 17 October 1965, without bothering to divorce his first wife, he married a "twenty-two-year-old former beauty queen."[8] This caused him to lose the support of the Catholic Church, which had until then been benevolent toward its former pupil. The president and his new wife returned from their Parisian and Brazilian honeymoon on 6 November. At the legislative elections held the next day (with only RDA candidates on the ticket), the population apparently marked its disapproval by staying away from the polls; the abstention rate was as high as 60 percent in Ouagadougou. Yet the single list still officially received 99.98 percent of the votes. A postelection government reshuffle in December marked a further increase in Yaméogo's personal power with the appointment of two of his relatives, Edouard and Denis Yaméogo, the latter with the interior portfolio. On 27 December, amid a burgeoning but already severe fiscal crisis and in an unfortunate contrast to the expensive glamour of the president's wedding and honeymoon, the new Yaméogo government announced a sharp austerity budget for 1966, with salary cuts of up to 20 percent in the civil service and a reduction of social security payments. The USTV initiated a meeting two days later to set up an interunion committee to oppose the new austerity measures.

On 31 December, following a heated meeting with Interior Minister Denis Yaméogo,[9] union leaders called for a general strike for 3 January. The same evening their offices were surrounded by troops firing tear gas canisters. On 1 January Yaméogo declared a state of emergency, arrested several of his leading opponents, and accused CATC's Joseph Ouédraogo of being an agent of China and of fomenting a "subversion of communist inspiration."[10] The strike went on as planned, despite the presence of troops in the streets. Far from intimidated by the troops, demonstrators actually chanted slogans demanding a military takeover. In the evening army chief of staff Colonel Sangoulé Lamizana took over at the unions' request, and Yaméogo resigned. On 4 January union leaders announced their support of the new regime.[11]

## Sangoulé Lamizana: The Benevolent Dictator

### Opening Pandora's Box: The First Military Regime

Despite the RDA's pervasive presence, its political hegemony had remained quite superficial throughout Yaméogo's rule. The regime's lack of consensual foundations made Yaméogo's presidency much more fragile than his overwhelming electoral success in October 1965 indicated. The speed with which an aspiring totalitarian regime such as Yaméogo's crumbled is astonishing and reveals deeper

*President Sangoulé Lamizana. Courtesy of* Sidwaya.

political instability. The enthusiasm with which the unions called on the military to take over is equally remarkable, especially in view of the number of politicians in their ranks who no doubt saw themselves fit to succeed the ousted president. It appears the union leaders were hoping the military would simply play a transitional role before elections would allow them to return to politics. They would have the next decades to learn that it is easier to bring the army into politics than it is to get it out.

A Muslim of Samo origin, President Lamizana was a career officer, a member of the French armed forces from 1936 to 1961 and a veteran of World War II and of the wars in Indochina and Algeria. In June 1966 he lifted the state of emergency and legalized political activities. But on 12 December 1966 a new ban was imposed on all activities of political parties following clashes between supporters of Joseph ki-Zerbo and Yaméogo in Koudougou. On the same occasion, while repeating that the return to a civilian regime remained the government's objective, the army announced that it would stay in power for a period of four years, stunning most politicians, who expected an earlier return to civilian politics. The organizational rights of trade unions were not affected, however.

The constitution had been suspended on 3 January 1966 to make room for new institutions: the Superior Council of the Armed Forces and the Consultative Committee. President Lamizana, in addition to being head of state and minister of defense (until 1967) also headed the Superior Council, to which all officers from the rank of captain and above belonged. More than half the government's ministers were also soldiers. As for the forty-six-member Consultative

Committee, it comprised ten officers, five trade union representatives, four delegates from political parties, ethnic chiefs, religious leaders, and members of cultural and professional associations. All its members were appointed by the military, making the Consultative Committee a figurehead institution. Despite a civilian presence, the first Lamizana regime thus remained overwhelmingly military in nature.

In 1969 Lamizana announced the termination of the military regime, the preparation of a draft constitution, and the resumption of political activities. The RDA resurfaced with two main leaders, Joseph Ouédraogo and Gérard Kango Ouédraogo, the latter formerly a member of the MDV and ambassador to London. Joseph ki-Zerbo's MLN also appeared somewhat popular among urban and university elites. In addition, Nazi Boni returned from exile to resurrect his PRA, but he soon died in an accident. Although it recognized political parties, the draft constitution provided for a four-year period during which the head of state would be the most senior military in the highest rank (i.e., none other than Lamizana himself) and during which the military would be represented in all political institutions. The new National Assembly was empowered to withdraw its confidence in the government, but the prime minister would still be appointed by the president. Despite its maintenance of military rule, the draft constitution was approved by referendum on 14 June 1970, with 98.41 percent of voters consenting and an abstention rate of about 25 percent. Upper Volta's "second republic" was thus ushered in.

## The Second Republic (1970–1974)

Following the approval of the new constitution, legislative elections were held in December 1970. Out of fifty-seven seats, the RDA won thirty-seven, the PRA secured twelve, the MLN six, and a dissident splinter group of the RDA called the Independents' Party two. In February 1971, from the ranks of the victorious RDA, Lamizana chose as his prime minister Gérard Kango Ouédraogo, who in turn appointed a government with eight RDA members, two PRA, and five officers. Joseph Ouédraogo, the RDA's secretary general and the CATC's president, became the National Assembly's speaker. Although formally allies in the same party, the two men—commonly referred to as Gérard and Joseph in Ouagadougou—shared a profound rivalry. Gérard, born in 1925 in Ouahigouya, was a *nakombse* of Yatenga, the uncle of the Yatenga *naba*.[12] Joseph was a Mossi of Ouagadougou, formerly a member of the *mogho-naba*-sponsored Union pour la Défense des Intérêts de la Haute-Volta. Their antagonism found roots both in the Yatenga-Ouagadougou friction and in the class difference between a member of the nobility and a commoner.

After almost two years of civil peace, union discontent resurfaced in January 1973 with strikes organized by the union of primary school teachers, the Syndicat National des Enseignants Africains de Haute-Volta (SNEAHV), and the union of secondary school and university educators, the Syndicat Unique Voltaïque des

Enseignants du Secondaire et du Supérieur (SUVESS). The unions demanded free housing, better promotion policies, and the Africanization of education ministry personnel, among whom expatriates were still numerous. The strike ended in February, but in April the Organisation Voltaïque des Syndicats Libres (OVSL), a new federation of unions, began a solidarity strike that turned to looting within days and was repressed. In April the MLN, which was then politically close to SNEAHV and SUVESS and a dissident faction of the PRA, followed up on the unions' action and called—without success—for popular mobilization against the government's "dictatorship."

The strikes of 1973 having weakened the authority of Gérard Kango Ouédraogo and his government and the deadline for the 1974 presidential elections drawing nearer, Joseph Ouédraogo made a move in December to get rid of the prime minister and to dismiss him from his party. With about twenty deputies loyal to his side of the RDA, Joseph Ouédraogo joined the opposition and created a new numerical majority that censored the government. Gérard Kango Ouédraogo refused to resign, and the stalemate brought all parliamentary and government work to a standstill. On 8 February 1974 Lamizana stepped in and suspended the constitution, dissolved the National Assembly, banned all political activities, and dismissed the Ouédraogo government before appointing a new one (with ten soldiers and four civilians) that he labeled the government of "*renouveau national*," or national renewal.

## The Failure of National Renewal (1974–1978)

The radicalism of Lamizana's reaction to the political deadlock betrayed his exasperation with civilian politicians and with multiparty parliamentary democracy. The events appeared to all but convince the president of the lack of maturity of Voltaic politicians. Searching for ways around the divisiveness of pluralistic institutions, he announced in February that "an institution for renewal will be set up in which all [political and social] forces contributing to the country's development will be represented."[13] In May the president outlawed political parties and called on "all the sons of the republic" to join the "movement for renewal."[14] Only unions and freedom of the press were maintained. Yet it was not until November 1975 that Lamizana announced the actual creation of a new single party, the Mouvement National pour le Renouveau (MNR), to be the only arena for all political, economic, cultural, and social activities. For lack of political parties, the reaction to this announcement came from the trade unions. Several strikes culminated in a two-day general strike on 17 and 18 December to demand the dissolution of the government and present the unions' program for a return to a constitutional civilian system. President Lamizana lacked both the conviction and the political support to impose his project. He dissolved his government on 29 January 1976 and shelved the MNR for good. A new government, appointed in February, was made up of two-thirds civilians and one-third members of the military. The unions' victory was overwhelming. A commission was set up to define the conditions of a return to a civilian regime; the president accepted its recommendations in November.

Having witnessed the failures of both political monopoly and competition since independence, Lamizana approved a compromise draft constitution that set up a system of political oligopoly, or limited competition. The draft provided that after a period during which any party would be granted a chance to draw supporters, only the three parties that won the most votes in the first legislative elections would be allowed to survive. The other parties would lose their legal existence and be forced to merge with one of the three official parties. On 27 November 1977 this constitution was adopted by referendum, with 92.7 perecent approval.

## *The Third Republic (1978–1980)*

Legislative elections were held on 30 April 1978 and two-round presidential elections on 14 and 28 May. Eight political parties, including the RDA and the PRA, registered for the legislative elections. Joseph ki-Zerbo's MLN had merged with smaller groups, and its candidates ran under the banner of the Union Progressiste Voltaïque (UPV). A fourth substantial contender was the newly created Union Nationale pour la Défense de la Démocratie (UNDD) chaired by Herman Yaméogo, the son of Maurice Yaméogo, and composed of many former members of the RDA, some of whom supported the dismissed president and others of whom had grown weary of the permanent infighting between "Joseph and Gérard." Only about half the eligible voters registered, and abstention reached 59.8 percent. No single party captured an absolute majority in the fifty-seven-seat assembly: The RDA won twenty-eight seats, the UNDD thirteen, the UPV nine, the PRA six, and the Union Nationale des Indépendants (UNI) one. In addition, the RDA's relative majority was further weakened after Joseph Ouédraogo took five other RDA deputies with him to create the dissident Front du Refus. Yet the PRA and UNI joined the twenty-two remaining RDA members to create a parliamentary majority (although they merged their deputies with the RDA, the two parties never officially dissolved, despite the constitution). The Front du Refus, the UNDD, and the UPV thus joined the opposition. Gérard Kango Ouédraogo was elected speaker of the assembly over Joseph Ouédraogo.

Because of the RDA split, four candidates ran in the presidential elections. The RDA, PRA, and UNI supported the candidacy of President Lamizana, who did not himself belong to any party. Because Herman Yaméogo was, according to the constitution, too young to be president, the UNDD's candidate was Macaire Ouédraogo, who made no secret that if elected he would reinstate Maurice Yaméogo in his political rights. Joseph ki-Zerbo was the UPV candidate and Joseph Ouédraogo ran for the Front du Refus (the "refusal" of his front was with respect to the RDA's support for Lamizana's candidacy). Abstentions reached a new record on 14 May, with 64.8 percent of registered voters staying at home, and Lamizana failed to get an absolute majority over runner-up Macaire Ouédraogo.

On 28 May, however, the outgoing president won with 56.2 percent of the votes to Ouédraogo's 43.8 percent; 56.5 percent of voters stayed away from the polls.

Lamizana appointed Joseph Conombo prime minister in an attempt to sideline both Joseph and Gérard Kango Ouédraogo. Conombo's government comprised nineteen civilians and two military men. The president himself had not resigned from the military. Eleven ministers were RDA[15] and four PRA. Despite Conombo's efforts to form a government of national unity, the UPV refused to leave the opposition and eventually merged with the Front du Refus into the Front Progressiste Voltaïque (FPV), thereby reinventing the usual bipolarity of Voltaic politics.

The third republic proved no more adequate than earlier regimes. The National Assembly remained virtually paralyzed and pointless. The government did not respond before the parliament to written questions from deputies.[16] In addition, in a similar pattern to that of 1973, Joseph Ouédraogo's FPV spent much political energy trying to topple the speaker of the house, Gérard Kango Ouédraogo. The political regime, self-centered and highly personalized, was thus ill prepared to cope with social conflict when it surfaced. And in true Voltaic fashion, it surfaced soon enough in the form of labor unrest. From 1 October to 22 November 1980 primary school teachers, unionized in the SNEAHV, launched a strike to demand higher professional status and protest alleged nepotism in the administration. Politically close to Joseph ki-Zerbo, the SNEAHV saw its action relayed to parliament by the FPV. Simultaneously, however, the communist-inspired Confédération Syndicale Voltaïque (CSV) and its leader, Soumane Touré, backed the strike with demonstrations in November. Alluding to the radical affiliation of Soumane Touré (a member of a communist organization called the Ligue Patriotique pour le Développement, or Lipad[17]), president Lamizana stigmatized the social and political upheaval as "external communist manipulations."[18] Yet despite union and political backing, the strike gained little popular support, as it kept children out of school for almost two months; the SNEAHV called an end to it on 22 November.

While the Conombo government appeared to have emerged victorious from its confrontation with the unions, the regime had in fact been weakened and was perceived as ripe for overthrow by some segments of the military, displeased as they were at the apparent spectacle that politicians had once again made of themselves. On 25 November soldiers and armored vehicles from the Régiment Inter-Armes d'Appui (RIA) invaded Ouagadougou under the orders of Colonel Saye Zerbo, who declared a state of emergency, suspended the constitution, banned political parties, and granted the SNEAHV its earlier demands. The surprising extent to which the population welcomed the suspension of its political freedoms revealed a high level of dissatisfaction with the Voltaic experience of parliamentary democracy. In addition to the usual rallies (some staged by the new government), backing came from other quarters: The *mogho naba* called the coup "an unhoped-for chance for the country and the people."[19] Catholic Archbishop

Zoungrana talked of "a blessing from God" and labeled Saye Zerbo "an agent of providence, of a God who loves Upper Volta."[20] And even former president Yaméogo expressed his support.[21] All those who had stood to lose or who felt sidelined by the civilian Lamizana regime—SNEAHV and the FPV, the Mossi (Lamizana was Bissa), the Catholics (Lamizana was Muslim), the UNDD—cheered the new government (although the new president, Saye Zerbo, was also a Muslim Bissa).[22] The unions' backing (another pattern reminiscent of 1966) was expressed on 1 December by the SNEAHV and the SUVESS, which declared themselves ready to take part in the "drive for recovery."[23]

## The Comité Militaire pour le Redressement et le Progrès National

### The Reinforcement of Praetorianism

Saye Zerbo, the son of a veteran, had been raised in a military environment. He went to primary school in Mali and secondary school in Senegal before heading for Paris for a degree in economic and social development and a staff college certificate from the military school. He had been Lamizana's foreign affairs minister from 1974 to 1976. To Lamizana's generation of officers, Saye Zerbo represented what technocrats are to traditional politicians: education over militancy, efficiency over politics, hierarchy over democracy.

Upon suspending the constitution, Saye Zerbo set up the Comité Militaire pour le Redressement et le Progrès National (CMRPN).[24] Unlike earlier military ruling bodies, the CMRPN extended its membership below the rank of captain; five noncommissioned officers and one rank-and-file soldier were added to its core group of twenty-five officers. Government portfolios were shared among eight members of the armed forces and nine civilians. None of the latter had ever exercised ministerial functions, and none was a politician in the sense of the third republic. The CMRPN perceived its mission as one of economic rigor and "purification" of the polity. Austerity measures were introduced in the civil service and in the military itself. Embezzlers of the third republic were prosecuted. Priority in economic development was given to the rural sector, as Saye Zerbo frequently toured the countryside. In Ouagadougou the CMRPN forced bars to close during business hours to combat absenteeism in the civil service, instituted an organ of censorship of the press, and banned the right to strike in November 1981 as the unions grew leery of austerity.

The latter measure drew different reactions. The CSV, which opposed it, was dissolved. The SNEAHV, for its part, seized this opportunity to distance itself from the CSV and did not confront the CMRPN. Nor did the USTV and OVSL. In fact, three of the civilian ministers in the Saye Zerbo government were teachers with links to the SNEAHV and the FPV. The SUVESS, however, had radicalized its leadership at its September 1981 congress and stood by the CSV and Lipad in their opposition to the strike ban and to the CMRPN. Yet the CMRPN's strong hand prevented the occurrence of a social confrontation similar to that of 1966 or 1980. Lacking an outlet in the social arena, however, political factionalism erupted at the very core of power: in the military.

## Upheaval and Politicization in the Military

The CMRPN differed from earlier military systems in its emerging factionalism. The first split was between senior and junior officers, veterans of World War II and products of military academies.[25] Although Lamizana had eventually joined the ranks of the RDA, lost his military referee role, and had thus been overthrown as a de facto civilian, he still symbolically represented the traditional national military. The other split was more conclusively political. While Lamizana had been politically evenhanded until 1978, keeping a balance at least among the RDA, the PRA, and the MLN/UPV, the CMRPN that succeeded him implicitly sided with the FPV and the SNEAHV and against the RDA and the CSV. By doing so, the CMRPN opposed both the traditional conservative forces and the emerging communist undercurrents in Voltaic politics, favoring instead the moderately progressive politics of Joseph ki-Zerbo. Those excluded—conservatives and radicals—resorted to alliances in the military to further their desire to return or advance to power. This they did by following Saye Zerbo's precedent, that is, without waiting for the explicit invitation of civil society.

## The Conseil de Salut du Peuple: A Prelude to the Revolution

The main consequence of the CMRPN's rule was thus the joint effect of increased militarization of politics and politicization of the military and the interruption of the cyclical returns to civilian rule that characterized Lamizana's government. By ruling out a civilian answer to Upper Volta's problems, Saye Zerbo's regime naturally led to a further fragmentation of the military, as only soldiers saw themselves capable of correcting the policies of other soldiers. This opened the way to the violent political instability of the 1980s, which witnessed successful coups in 1982, 1983, and 1987; a military reshuffle in 1983; and alleged (and violently repressed) unsuccessful coups in 1984, 1985, and 1989.

The overthrow of Saye Zerbo on 7 November 1982 revealed profound divisions and antagonisms in the armed forces. The coup was a far from homogeneous move and actually brought contending factions to the political forefront. As the future president, Thomas Sankara, later acknowledged, it was "a very complex story . . . [and] gave birth to a very heterogeneous and composite power, with its inevitable contradictions."[26] Apparently, the prime responsibility for the coup belonged to a leader of the military's conservative faction, Commandant Gabriel Somé Yorian, whose aim may have been to arrange for Yaméogo's return to politics. Yet he received the support of the army's other main political bloc, the radical faction, which was anxious to get rid of Saye Zerbo because of his repression of the CSV and Lipad. Under the leadership of Captains Thomas Sankara and Blaise Compaoré, this faction was specifically opposed to the anti-union stance of the regime and generally eager to launch a revolution and install a "popular" democracy. Because of their sharply different agendas, the putschists were unable to agree on a future leader from their own ranks and eventually settled on a low-profile figurehead, Médecin-Commandant Jean-Baptiste Ouédraogo, a military physician without experience in a position of actual command. Behind the powerless Ouédraogo, however, the Conseil de Salut du Peuple (Council for the People's Salvation, or CSP) was actually a collegial ruling body, though one ripe with dissension. The multiplicity of factions swelled its ranks to 120 members—forty officers, forty noncommissioned officers, and forty privates—marking an acceleration of the "juniorization" of military power begun under Saye Zerbo.

The first weeks witnessed the influence of Somé Yorian's agenda. The new president announced on 17 November that a new "normal constitutional life" would be resumed after a reorganization of the army and of the state apparatus. Union and press freedoms were reinstated. Soon, however, the Sankara-Compaoré faction gained some popular ascendancy and managed to have Sankara appointed prime minister, creating a de facto counterpower to Jean-Baptiste Ouédraogo and highlighting the CSP's political ambiguity. Whatever agreement may have existed between Somé Yorian and Sankara, the latter ignored it as soon as he was appointed prime minister and immediately used his position to further the views and policies of his group and to foster his own popular support. In just a few months, he redirected the country's traditionally conservative foreign policy toward alliances and friendships with Algeria, Ghana, Benin, Cuba, Libya, and North Korea (the latter two of which Sankara visited as prime minister, apparently without Ouédraogo's knowledge). On the domestic front Sankara alienated the conservative forces in the military by stigmatizing politicians and those he labeled "bourgeois" as enemies of the people and by claiming that the army was to stay in power. In his inflammatory "speech of the truth" of April 1983, he declared, "The army wants power and democracy. It really wants to merge with the people. It is the army of the people. . . . We will hunt down rotten soldiers. . . . If tomorrow we could transform Upper Volta as Qadhafi transformed Libya, would you be happy?"[27] Following his prime minister's speech, President Ouédraogo repeated his goal of "true democracy" and a return to "normal constitutional life," warning Sankara that "it is out of the question to impose an ideology or a model of society [in Upper Volta]."[28]

After Sankara delivered a similar speech in Bobo-Dioulasso on 15 May, the conservative wing of the CSP decided to remove him from office and arrested him on 17 May. Compaoré escaped a similar fate, managing to reach his Pô garrison, which supported him against the CSP. In Ouagadougou a new government—labeled CSP-2—was appointed without representatives of the radical faction. President Ouédraogo made no secret of the reasons behind Sankara's eviction: "It is a problem of ideology.... We were following step by step the program of the [Lipad], and that program was to lead us to a communist society."[29]

Yet stability was as elusive for the CSP-2 as for its predecessor. On 20 May 1983, high school and university students, manipulated by Lipad, demonstrated violently in the streets of Ouagadougou, demanding the reinstallment of Sankara. Meanwhile, after his arrival in Pô, Compaoré had begun an open military rebellion against the CSP-2. Finally, it appears that their ouster helped the radicals momentarily to forget their differences and unite in a more influential bloc in contact with Compaoré.[30]

Because of the ongoing crisis, Ouédraogo hastened the implementation of his program but could not avoid making political blunders. His rehabilitation of Maurice Yaméogo, for example, angered politicians Yaméogo had sidelined in the early 1960s and who were forbidden to return to politics under the rules of the CSP-2. Ouédraogo went on with the release of politicians from the third republic and eventually succumbed to pressures to free Sankara himself. Although the CSP-2 had also banned members of the armed forces from political activities (despite being itself a military regime), Sankara used his relative freedom (he was still under house arrest) to negotiate a government with the radical civilian political groupings and to arrange a takeover with Compaoré.

## The 1983 Coup and the Conseil National de la Révolution

With Sankara under house arrest, it was Blaise Compaoré who took the initiative of the coup from his military stronghold in Pô near the Ghanaian border. Having seized the vehicles of Canadian development workers, Compaoré's combat-ready men headed 90 miles north to Ouagadougou on the evening of 4 August, freed Sankara, and captured the national radio where Sankara delivered a short communication proclaiming the revolution, announcing that the Conseil National de la Révolution (CNR) had taken over, and inviting the population to form neighborhood Comités de Défense de la Revolution (CDRs). Jean-Baptiste Ouédraogo surrendered, and Gabriel Somé Yorian was executed. This was the first coup in Upper Volta's history that resulted in casualties.

In its first year the CNR was composed of four military leaders—Captains Thomas Sankara, Blaise Compaoré, and Henri Zongo and Major Jean-Baptiste Boukary Lingani—and of two civilian groups—the Lipad and the Union des Luttes Communistes (ULC).[31] Lipad, a branch of the Parti Africain pour l'Indépendance (PAI), an embryonic transnational West African communist party, was (and still is) an urban-oriented Marxist-Leninist formation. A self-labeled elitist vanguard party, Lipad was politically linked to the CSV, which gave it greater capacity to mobilize. CSV chairman Soumane Touré was also leader of the Ouagadougou section of Lipad. The ULC had been created in 1978, at the same time as the Parti Communiste Révolutionnaire Voltaïque (PCRV). Both the ULC and PCRV were the result of a schism in the Organisation Communiste

Voltaïque,[32] a matter of differences on obscure political points and personal quarrels. The PCRV refused to participate in the CNR, claiming it was not a truly revolutionary movement.

## *The Man and the Revolution: Thomas Sankara*

Thomas Sankara was a fairly nonconformist and charismatic leader. Before he seized power at the age of thirty-three, he had been commander of the Pô paratroopers, reputedly the most efficient unit of the Burkinabè army. He became popular by having his men help local peasants harvest and by forming a band, the Pô Missiles, that performed at local gatherings (Sankara playing the guitar). Later, as head of state, he held jam sessions with Ghana's president Jerry Rawlings during a public meeting and with British aid activist and rock singer Bob Geldof.[33] His unit also enlivened traditional military parades by presenting acrobatic exercises on motorbikes on Independence Day. In 1974 he fought the border "war" with Mali and returned home a hero.

A Silmi-Mossi—the son of a Mossi mother and a Peul father—Sankara personified the diversity of the Burkinabè people. His father, born in 1919, was a conscript in World War II who was taken prisoner in Germany in 1940; he later became a postal clerk in the colonial administration.[34] Born in Yako in 1949, Sankara was the third of ten children and the first of four boys. Because the family moved as his father was transferred, Sankara lived in Ouagadougou until the age of six, then in Gaoua and Bobo-Dioulasso, among other places. This, along with his later military assignments, gave him a solid background knowledge of his country. At age seventeen he was admitted to the military cadet school of Kadiogo, from which he received his baccalaureate. His history professor was Adama Touré, a Lipad leader who would be information minister in the first CNR government and who is believed to have been influential in Sankara's Marxist conversion. From 1970 to 1973 he was sent to the military academy of Antsirabe in Madagascar to be trained as an officer. He then studied in France (where he met civilian students who would later be the core of Burkina's leftist organizations) and in Morocco. As a minister of information under Saye Zerbo's rule, he went to his own swearing-in ceremony on a bicycle and resigned after a few months, cursing "those who gag their people."[35] Once president, he auctioned off the government's fleet of Mercedes vehicles and made the small Renault 5 the official government car. On his way to summits of the Organization of African Unity (OAU) and other international conferences, he regularly hitched rides on the planes of neighboring heads of state.

A good though long-winded orator, he conversed with his audience rather than addressing it. His speeches, all long and many improvised, were didactic and imaginative.[36] René Otayek ably expresses how, for many young people in Africa, Sankara not only represented the rejection of their lot but also brought hope, energy, and another realm of possibility and pride.[37] To this day he remains a powerful symbol among Africa's youth. Most visitors who met with the president

*President Thomas Sankara (with walkie-talkie) and bodyguards. Courtesy of* Sidwaya.

agreed that he had an uncommon vitality and dedication to his office. He was extremely likable, and his warm honesty for a time prevented many from seeing the shortcomings of his policies.

## Ideological Confusion

Though for the most part a populist regime, the CNR relied on a relatively wide array of radical ideologies to legitimate its power. The few official attempts at placing the revolution within an ideological model were at best confused, but it is clear that its populism was accompanied by elements of Marxism-Leninism and nationalism. But the driving ideological force behind the revolution was Sankara's sense of "Robin Hoodism," of social justice, redistribution, and sharing. Yet Sankara's ideological and conceptual acumen was inadequate, and his indoctrination by radical fellow revolutionaries led him to misread his country's conditions.

The regime's main ideological manifesto was the *Discours d'orientation politique (DOP)*,[38] a speech Sankara read on national radio about two months after taking office and widely distributed thereafter by all organs of the state and the CNR. The *DOP* highjacked Voltaic history for the benefit of the CNR, which was represented as the historical revolutionary outcome of social conflicts since independence. In doing so, it divided Voltaic society into classes aggregated as either "the people" or "the enemies of the people" according to their supposed morality rather than objective conditions.[39] The people were the working class, the petite bourgeoisie, the peasantry (categorized as part of the petite bourgeoisie), and the lumpen proletariat (the unemployed underclass). As for the enemies of the people, they were the bourgeoisie—split into state bourgeoisie (civil servants and the traditional political class), comprador bourgeoisie (businesses and traders), and middle bourgeoisie (undefined)—and the "reactionary feudal forces" (ethnic structures and chiefdoms).

Although most of the revolutionary leaders were civil servants and thus belonged to the state bourgeoisie—which should have made them enemies of the people—they considered themselves part of the petite bourgeoisie. They believed the latter had been called upon historically to lead the revolution but would be replaced in later stages by the working class and the peasantry.[40] The bourgeois enemies of the people were linked to "imperialistic" forces of which they were the local agents of exploitation, a pattern reminiscent of dependency theories. The CNR's stratification had unintended ethnic consequences. As the comprador bourgeoisie was essentially made up of Yarse (Mossi-assimilated traders) and Dioula, the latter two groups necessarily became enemies of the people and thus of the revolution.[41] And by accusing the Mossi chiefdoms of feudalism, the CNR's ideology had all but declared war on ethnicity. The peasantry (the people) was thus artificially severed from its own social structures (the enemies of the people). Furthermore, despite the near nonexistence of an industrialized working class in Burkina and the widespread peasantry, it was the workers rather than the peasants who were considered the real revolutionaries.[42] In the end the CNR's class analysis was the rigid imposition upon Burkina's society of a categorization designed for different places and different times. Its own ideological confusion would later cost the CNR its very survival.

Other features of the CNR's ideology were its anti-imperialism (targeted primarily at France, for historical reasons, and at the United States, for Burkina's leaders identified with those of Cuba and Nicaragua), its own assimilation to a movement of national liberation (witness the national motto, "Fatherland or death, we will triumph," borrowed from Cuba's "*Patria o muerte, venceremos*"), and its nationalism. The latter point is illustrated by the change of name on the first anniversary of the revolution: On 4 August 1984 Upper Volta (the name referring to the country's situation along the northern part of the Volta Rivers) became Burkina Faso.[43]

## *The Drive Toward Totalitarianism*

The CNR brought major changes to Burkina's political, economic, and social systems. Despite claiming to belong to the people, however, it was often at odds with society and fell into a spiral of increased totalitarianism and violence.

One of the new regime's very first actions was to set up the CDRs, composed mainly of young people, sometimes teenagers, many unemployed and eager for a role. Established in every village, neighborhood, town, office, ministry, and even the military, they held a contradictory position, as the government wanted them to be both an instrument of its control and "the local organs by which the people exerts its power."[44] They would prove better at the former than the latter, for they soon turned into a relay for the CNR, the expression of its power down to

*Political Instability and the Quest for Legitimacy* 59

the smallest structures of social life. It was a one-way street. Feedback from "the masses" was not tolerated and was prevented by systematic "self-critiques" at every level. In addition, other CNR-inspired organizations, such as the Union Nationale des Anciens du Burkina (UNAB), the Union des Femmes du Burkina (UFB) and the Mouvement des Pionniers—respectively, for elders, women, and children—left no social stratum untouched.

Although deprived of political autonomy, the CDRs also tended to wield a power of their own and soon evolved into a repressive structure. As early as April 1986, Sankara had to denounce their abuses and acknowledged instances of looting, torture, and embezzlement.[45] Yet he continued to rely on them to implement his policies. The CDRs' oppression was thus twofold: It was the expression of near totalitarian control by the CNR over every sector of society[46] and it was the fruit of some of its members' excesses. The use of the CDRs as instruments of repression, control, and mobilization made them deeply unpopular.

As often in Burkina's history, trade unions opposed the regime's totalitarian tendencies and the CDRs' encroachment on their traditional prerogatives of mobilizing and representing workers. The SNEAHV immediately called on its members to stay away from the new power structures. After eight months of tense relations, three of SNEAHV's leaders were arrested in March 1984 for "plotting with the FPV against the security of the state."[47] The arrests triggered a protest strike by SNEAHV teachers two weeks later. The CNR's response was extreme: Despite the government's recognition of the right to strike, it laid off between 1,500 and 2,400 teachers, about half the SNEAHV'S membership. The union was effectively wiped out. The CNR then encouraged the election of new union leaders who led SNEAHV, relabeled the Syndicat National des Enseignants du Burkina (SNEB), toward a prorevolution stance.

The CSV's attitude evolved differently. With the initial participation of Lipad in the CNR, the CSV was at first docile and refused to see a contradiction between its role and that of the CDRs. Yet after Lipad was ousted in 1984, the CSV, by then renamed the Confédération Syndicale Burkinabè (CSB), joined the opposition and fiercely fought the CDRs' growing influence. In early 1985 Soumane Touré was arrested for having accused CNR members of corruption. Other trade unions linked to the PCRV[48] also opposed the antiunion policy of the CNR. Like Yaméogo before him, Sankara tried to merge all unions into one organization but without success. Having failed to co-opt them, his regime then grew increasingly hostile toward them. Union members who in April 1987 had published a memorandum critical of the CNR's socioeconomic policies were arrested and detained without charge. Soumane Touré was arrested in May—for the third time since the beginning of the revolution in 1983—by his local CDR in Ouagadougou for alleged "subversive actions." Thus the CNR, like all previous regimes unable to swallow the unions, gradually crushed their power as no regime had before.

The Tribunaux Populaires de la Révolution (TPR) were other means the CNR used in taming Burkinabè society and, in Otayek's words, "dismantling the clien telist networks" of previous regimes, leaving the CNR "master of the political field."[49] Created in October 1983, the TPRs held their first session in January 1984. They were empowered to hear cases related to political crimes and misdemeanors, crimes against the internal and external security of the state, and embezzlement and other crimes committed by civil servants in the performance of their duties.[50] They were composed of five CDR members, one soldier, and a presiding magistrate. There was no public prosecutor, nor was the accused allowed a lawyer, as the regime feared lawyers' capacity to "transform truth in lies and lies in truth."[51]

The TPRs could impose sentences ranging from fines and expropriation to up to fifteen years of imprisonment. Most politicians from previous regimes were called before the courts. Lamizana was acquitted of any wrongdoing, but Gérard Kango Ouédraogo received a ten-year sentence (of which he served two), Saye Zerbo a fifteen-year sentence (later commuted to house arrest), and Joseph ki-Zerbo was condemned in absentia to two years in jail and several million CFA francs in fines. With the TPRs the CNR made itself the judge of history and encouraged submission to its new order. But while sessions of the TPRs were nationally broadcast, seven alleged attempted coup plotters were executed without proper hearings in 1984, and prisoners were allegedly tortured after Ouagadougou was rocked by bomb explosions in May 1985. The CNR thus did not hesitate to use raw violence when it believed its stability was threatened.

The CNR also used other methods to subdue citizens. It imposed a one-year military and civil service on everyone between the ages of twenty and thirty-five; civil servants were required to work their ministry's "collective fields" and attend weekly sports sessions. In January 1987 the CNR initiated a compulsory dress code; all government officials were to wear the national costume, the *faso dan fani*. Similar arbitrary measures were intended to force popular mobilization.

## Political Disorder and the Collapse of the CNR

During the first year of the revolution, Lipad had the largest representation in the CNR. In August 1984, however, after Lipad had increasingly attempted to turn the revolution in its own ideological direction and to infiltrate all organs of power, its members were ousted from the CNR, the government, and most positions of responsibility in an episode known as the "clarification."[52] This brought the ULC into a more influential position, as it became the strongest civilian element to face the Organisation Militaire Révolutionnaire (OMR), as the radical officers labeled themselves. Yet the price of "clarification" was a reduction in the social basis of the CNR and an increase in political opposition to the extent that Lipad reactivated its unionist branch, the CSB, against the regime.

The year 1985 was therefore difficult for the CNR, which relied more and more on violence, constraint, and repression to stay in power. In the course of 1986, however, additional groups were incorporated into the regime or given more room in an attempt to enlarge its political foundations. The two parties that benefited from this policy were the Union des Communistes Burkinabè (UCB) and the

Groupe Communiste Burkinabè (GCB). The UCB was set up by Sankara himself in order to check the influence of the ULC and was headed by the secretary general of the CDRs, Lieutenant Pierre Ouédraogo.[53] It was not believed to act independently from the OMR. The GCB, in contrast, created in 1983 as a dissident faction of the PCRV, was already in the CNR. Because it was numerically a very small political group, it contributed little to the attempt to broaden the regime's basis.[54]

By the beginning of 1987, the GCB, the UCB, and the ULCR were sharing power with the military. In many ways Sankara had played a balancing role among all these factions over the years, but as of 1987 he had begun to understand his power in increasingly personal rather than collegial terms. Having earlier lost control of the UCB, however, he lacked some organizational basis to support his influence as president. To circumvent this weakness, he acted both on the political and military fronts. He first pushed for the creation of a single party to merge all CNR groupings,[55] a suggestion poorly received by the other factions, who feared for their independence, their power, and their careers. His second move backfired as well, this time alienating his military friends. Unlike those CNR officers, Sankara had no direct military command, so he proposed the creation of a paramilitary force to be called Force d'Intervention du Ministère de l'Administration Territoriale et de la Sécurité (FIMATS) and to be headed by Vincent Sigué, an author of the 1983 coup who was reputed to be responsible for opponents' torture under the CNR.

Both moves were rejected by the other CNR members. By then Sankara's power was so marginalized that he was no longer in actual control of the state. It appears in retrospect that he could simply have been sidelined by his opponents. Yet on 15 October 1987 he was ambushed and assassinated at his office by soldiers belonging to Compaoré's entourage. Whatever the legitimacy of Compaoré's claim that he was unaware of this initiative and thus had not authorized it, the coup translated Sankara's political marginalization into his physical elimination. The coup was extremely violent by Burkinabè standards: Sankara and at least twelve aides were killed, and others were later imprisoned, tortured, and exiled.

With the CNR deprived of Sankara's balancing function, the coup also cleared the way for increased control of the state by Compaoré, the UCB, and the GCB, acting to get rid of the ULCR. At stake was power over the left wing of Burkina's political spectrum. Sankara had considered ULCR's Valère Somé to head his proposed single party; many ULCR members went into exile in the wake of Sankara's murder. Sankara's failure to institutionalize his power (and the CNR's parties' failure to tame their ideological disputes) was thus central to the breakdown of his regime, its fate ultimately linked to that of his person.

## The "Rectification" and the Front Populaire

The Front Populaire first appeared to represent the continuation of the CNR, at least by the similarities in personnel and political alliances. When he took over,

Compaoré stipulated that the revolution was still on the agenda, although it had entered a process of "rectification." But the regime's political direction increasingly tilted away from the revolution, without ever effecting a clean break. As of early 1995 the entourage of the president was still composed of former orthodox revolutionaries, such as Arsène Yé Bognessan, the parliament's chairman. What has occurred is an inflow of other leaders onto the political scene without an eviction of the radicals, who have altered their stated position without clearly rejecting earlier policies.

## *The Choice of Violence*

Despite Sankara's dwindling popularity at the time of his death, Compaoré's violent accession to power generated more hostility than enthusiasm. Special missions had to be sent to the countryside to explain the events. Because of student protests, schools were closed nationwide for a week. Pro-FP demonstrations were boycotted, and mourners kept constant vigil at Sankara's makeshift grave. In an effort to boost its own credibility, the FP denigrated Sankara and persecuted his widow, Mariam, depriving her of employment and preventing her from leaving the country. In January 1988 it called a national conference in which 1,500 delegates from the thirty provinces of the country appraised government under Sankara. At the same time the FP adopted economic measures popular with civil servants and urban dwellers, such as the release of imprisoned union leaders, the lifting of a ban Sankara had imposed earlier in the year on imports of fruits and vegetables, a reduction in the price of beer, and raises of 4 to 8 percent in civil servants' salaries.[56]

Compaoré also promptly acted to sterilize political opposition. Following the announcement of his coup, a pro-Sankara unit of the military based in Koudougou and led by Lieutenant Boukary Kaboré initiated an armed rebellion. It was put down by Compaoré's troops. Kaboré fled to Ghana, and nineteen of his fellow rebels were executed. As for the ULCR, most of its leaders were arrested. According to *West Africa*, twenty soldiers and twenty-two civilians were detained as of February 1988; the Association des Juristes Africains counted thirty political prisoners.[57] Later in 1988, however, most detainees who had not been executed were released, and Mariam Sankara was allowed to leave the country for Gabon. In May Sankara's remains were buried in an official ceremony.

Compaoré is the opposite of Sankara. Shy and lacking Sankara's communication skills, he first appeared ill at ease with the function of president. And while Sankara was Spartan and austere, Compaoré enjoys luxury and is less strict on ethical matters. The compact-sized cars used in Sankara's administration have disappeared. A tall and muscular Mossi who married the niece of Côte d'Ivoire's late president Houphouët-Boigny, Compaoré is often called *le beau Blaise*, "handsome Blaise." He met Sankara in school, and they were inseparable until the overthrow. Ever since 15 October 1987, Compaoré has claimed that he did not want to kill Sankara and that he broke into tears when he saw his friend dead. He did not appear in public for four days after Sankara's murder. Yet he later actively criticized the former president and accused Sankara of having planned to kill him the very same day.

In its statutes published in March 1988, the FP described itself as a "grouping of political organizations and anti-imperialist and democratic mass organizations."[58] It was composed of a congress, a Coordinating Committee, and an Executive Committee. The UCB had the most influence within the FP, but the GCB and a faction of the ULC headed by Kader Cissé (president of the Conseil Révolutionnaire Economique et Social under Sankara) were also present. Also in March 1988 the CDRs were dissolved and replaced with Comités Révolutionnaires, CRs, which, however, never took off as credible institutions and never succeeded in imposing the same pressure for mobilization as had their predecessors.

The FP also took some ideological distance from the CNR. In October 1988 Compaoré boldly stated in an interview to *Le Monde* that "socialism had never been on [Burkina's] agenda."[59] Yet the FP's progressive restructuring of the state did not prevent the continuation of military factionalism. Captain Guy Lamoussa Saygo, a pro-Compaoré officer who had crushed the Koudougou rebellion in October 1987, was assassinated in his sleep at his house in Bobo-Dioulasso in November 1988 when a hand grenade was thrown through the window of his bedroom, also killing his wife. The FP's response was immediate and sharp. Seven rank-and-file soldiers from the Koudougou battalion—all supporters of the exiled Boukary Kaboré—were found guilty of the crime, sentenced to death on 28 December, and executed the next day.

On 15 April 1989 the government announced the creation of the Organisation pour la Démocratie Populaire/Mouvement du Travail (ODP/MT). According to officials, it was intended to rally all the revolutionary tendencies and groups belonging to the FP. The UCB and a contingent of the faction of the ULC that participated in the FP announced their self-dissolution and merger with the new party. Ten days later a major government reshuffle saw the ousting of the GCB and the ULC faction that had not agreed with the constitution of the ODP/MT. Clément Oumarou Ouédraogo, head of the UCB, became minister of coordination of the FP. Yet another "clarification" had taken place, and the regime now relied on an organization smaller than any other before. The April reshuffle marked the return of the single-party question to center stage. Yesterday's opponents had become today's proponents, having eliminated the danger of a ULCR-controlled single party.

The political gains the ODP/MT accumulated after the April 1989 government reorganization were confirmed during the first session of the FP's Coordinating Committee in June, which revealed the positions of influence occupied by some of its members: Clément Oumarou Ouédraogo was secretary for political affairs, and Captain Arsène Yé Bognessan (the head of the CRs) was secretary for organization. Yet despite its rising authority, the ODP/MT was still far from its desired hegemonic position as a single party. The 251 delegates present at the meeting represented three other political parties and four national unions. One of these three parties was the Mouvement des Démocrates Progressistes (MDP), created in October 1987 by Herman Yaméogo, the son of former president Maurice Yaméogo. Its presence marked the return of nonrevolutionary politics to Burkina Faso.

Another hindrance to the ODP/MT's ascendancy was the enduring power of Captain Henri Zongo and Major Jean-Baptiste Boukary Lingani, guardians of the integrity of the revolution as defined in 1983 and, as such, obstacles to political change. This hurdle was cleared in dramatic fashion on 20 September 1989 with the execution of Zongo and Lingani after the two men were summarily judged by a military tribunal and found guilty of plotting to overthrow Compaoré. According to Clément Ouédraogo and Arsène Yé, who announced the executions on national radio—Compaoré once again remaining aloof from the killings of former allies and friends—the two alleged plotters were going to attack the presidential plane as it brought Compaoré back from an Asian trip on 18 September. After having supposedly tried to enlist the support of the military hierarchy, the two men were betrayed by Captain Gilbert Diendéré, the FP's secretary for security. In Yé's words,

> in the president's absence, some important problems were referred to [Lingani] because he was the acting chairman ... and a number of contacts he had with certain comrades generated in him ambitions that have nothing to do with the highest interests of the revolution.... On the 15th Lingani summoned Diendéré to talk to him about what he described as the catastrophic national situation and asked that Diendéré help him so that, together, they could overthrow the FP.... In such a situation Diendéré could not refuse. So he agreed ... because his life was threatened ... and preferred to wait until [the 18th] to make his decision on taking the offensive.... [He called in] the chiefs of the fifth military region ... arrested Lingani's bodyguards but did not succeed in laying hands on Major Lingani, who had fled ... through the window of his office.[60]

It was only a short respite for Lingani, who was later arrested. Zongo went to the airport to welcome the president and attended a subsequent cabinet meeting before he, too, was arrested on the basis of allegations by Diendéré that Lingani had involved him in his plans. Tape-recorded confessions broadcast on 20 September ironically gave some credibility to the widespread theory in Ouagadougou that the two men had been framed. Lingani, though he recognized that the "rightist deviationism" of the revolution necessitated action, claimed to have been the victim of "manipulation." On 21 September an extraordinary session of the Coordinating Committee of the FP was convened in Ouagadougou and endorsed changes in the membership of the Executive Committee and a limited reshuffle of the government. Although his formal assignment did not change, Diendéré's influence within the military and the FP increased considerably after the September events. Diendéré had by then succeeded Compaoré as chief of the Pô commando garrison and was rumored to be presidential material. A Mossi from Fada N'Gourma, he was then only thirty years old.

## Calling the Shots: Compaoré's Political Maneuvering

The FP held its first congress in March 1990 with more than 3,000 delegates representing seven political parties, ten trade unions and nongovernmental organizations, the CRs, and provincial administrations. In addition to the ODP/MT and the MDP, other parties were the Groupe des Démocrates Progressistes (GDP), the Groupe des Démocrates Révolutionnaires (GDR), the Convention Nationale des Patriotes Progressistes (CNPP), and the Union des Démocrates et Patriotes du Burkina (UDPB). All these new parties—whose names made few references to radical ideologies—indicated the FP's continued drift away from the revolution and Compaoré's desire to expand its political scope. Yet control remained in the hands of the ODP/MT.

The congress marked a new turn in Burkina's cycle of successive authoritarianism and participation as it agreed to give the country a new constitution to lead toward "democratization." A constitutional commission was set up, but Compaoré—not unlike General Lamizana before him—reminded participants to the congress that if "Burkina's development cannot presently occur in a one-party system," nor could it afford "anarchic multipartism, which breeds national division." Apparently to appease the fears of ODP/MT members, he added that the constitution would "clearly define a socialist regime."[61] Compaoré's constitutional problem exemplified the dilemma in which the political gains of the ODP/MT in 1989 and early 1990 had put him. Though he appeared keen on opening up to more moderate forces, he needed to reassure the ODP/MT of his revolutionary and socialist commitment, as the latter maintained its stronghold over the FP. But Compaoré soon changed strategy and opted for another act of "clarification," ousting Clément Ouédraogo from the government and from the Executive Committee in April 1990 and replacing him with Roch Marc Christian Kaboré, formerly transport and communications minister, believed to be more accommodating to Compaoré's desire for political liberalization. In September 1990 Kaboré was appointed minister of state without portfolio.

Yet Compaoré's political liberalization remained limited and tolerated only among certain segments of the population. On campus, repression remained the order of the day. University lecturer (and ULCR member) Guillaume Sessouma had died in jail in December 1989. Demonstrations occurred on Ouagadougou's campus in May 1990 after ten students were expelled for voicing their grievances. The police moved in and arrested some thirty more. In October 1990 the student Boukary Dabo, arrested in May, was reported to have died in detention, too. Until early 1991 at least sixteen more students remained incommunicado in prison and were most likely tortured, while others had been conscripted into the armed forces.

The draft constitution delivered by the FP's constitutional commission in October had more to please the opposition than the ODP/MT. It provided for a multiparty system and the election of a National Assembly (for a four-year term) and president (for a seven-year term) by direct universal suffrage. The prime minister would be appointed by the head of state but responsible before the parliament. Social and individual rights would be recognized, and the Burkinabè society was described as "progressive, pluralistic and without prejudice." The draft also prohibited torture, declared men and women equal, and gave full freedom to the press. About 2,400 delegates, representing political parties, unions, CRs, traditional structures such as *chefferies,* and religious authorities, met in December 1990 to approve the draft constitution and further scrapped all references to "anti-imperialistic" and "revolutionary" from the definition of the state, calling it instead "democratic, unitary, and lay." A national referendum on the constitution was set up for 2 June 1991.

With the apparent beginning of a new era of presidential and parliamentary democracy, the ODP/MT dropped Marxism-Leninism as its official ideology at its first congress in March and proposed Compaoré's candidacy for the forthcoming presidential elections. In April 1991 the government announced amnesty for the remaining detainees arrested in connection with the attempted coup of December 1989, and Boukary Kaboré returned to Ouagadougou from his Ghanaian exile. Amid national campaigning for the referendum, the FP held an extraordinary congress in Ouagadougou on 24 and 25 May that approved its separation from the state following the referendum, called on political exiles to return home, and restored to Maurice Yaméogo his political rights, indicating Compaoré's affinity with the old system.

In the June referendum 93 percent of the voters approved the constitution. Yet 51 percent of the registered voters did not bother to show up at the polling stations, maybe in part because of the beginning of the agricultural season (participation rates were around 40 percent in most rural areas, as against 75 percent in Ouagadougou).

On 16 June Compaoré appointed a new government to run the country until the legislative and presidential elections planned for November and December. The reshuffle was substantial, as Compaoré intended to give at least the appearance of a government of national unity, despite the refusal of several parties to participate under his conditions. Indeed when he had dissolved the previous government on 12 June, Compaoré had convened a roundtable of all political parties to discuss the process of transition and constitutional change. Of twenty-four parties present at the onset of the roundtable, thirteen withdrew after Compaoré refused to turn it into a national conference similar to those that other African countries had been having to mark the change from authoritarian regimes to pluralistic ones since Benin began the tide in February 1990. The government created in the June reshuffle remained dominated by the ODP/MT, which counted no less than twenty-one ministers (62 percent of the government).

## The Fourth Republic: An Iron Fist in a Velvet Glove

A new government was appointed in late July 1991 to embrace representatives of more organizations outside the FP, including Herman Yaméogo as agriculture

*President Blaise Compaoré courts the favor of a Mossi chief during the 1991 presidential campaign. Courtesy of* Sidwaya.

minister. Yet in August Yaméogo and two other ministers from his party, renamed the Alliance pour la Démocratie et la Fédération (ADF), resigned over their enduring conflict with Compaoré about a national conference. Other opposition parties, such as the Alliance pour la Démocratie et l'Emancipation Sociale (ADES) and the Convention Nationale des Patriotes Progressistes/Parti Social Démocrate (CNPP/PSD), joined the ADF in calling for a sovereign national conference in August. In September about twenty opposition parties united around the ADF, the CNPP/PSD, and the ADES to form a Coordination des Forces Démocratiques (CFD) and gave Compaoré until 25 September to call a national conference. The deadline came and went, ignored by Compaoré, who had meanwhile announced his candidacy to the presidency. CFD ministers resigned in protest. In the following days clashes between CFD and government supporters ensued, and the ADF's headquarters were set on fire. Several demonstrations and counterdemonstrations were held in October; those *against* the national conference were organized by the Alliance pour le Respect et la Défense de la Constitution (ARDC), a new government-sponsored organization allegedly grouping together twenty-seven parties. After acts of looting and violence, the government imposed a ban on public demonstrations on 1 November.

In September, after eight years of exile, Joseph ki-Zerbo returned to Ouagadougou to attend the first congress of the CNPP/PSD. The group nominated Pierre Claver Damiba, the UNDP regional director for Africa, to be its candidate for the presidential elections. The RDA (a member of the CFD) chose Gérard Kango Ouédraogo as its presidential candidate. In October the elections were postponed to December and January for lack of consensus on the transitional period. But as the government continued to ignore their calls for a national conference, opposition leaders withdrew their candidacy and called for an election boycott. On 1 December 1991 Compaoré, the sole contender, won the presidential elections with 86.19 percent of the votes, amid a record 74.7 percent abstention rate and scattered violence.

Compaoré's success exacerbated the prevailing political tension, and on 9 December the secretary general of the Parti du Travail Burkinabè (PTB) and former chair of the ODP/MT, Clément Oumarou Ouédraogo, was assassinated. Other opposition figures were attacked the same night. The CFD held the government responsible for the murder, accusing it of getting rid of individuals who could make revelations damaging to the government were a national conference ever held. In an attempt to appease the opposition, the government announced the postponement of legislative elections and the rehabilitation of more than 4,000 people punished for political or trade union activities since August 1983.

On 24 December 1991 Compaoré was sworn in as the first president of the fourth republic. After further failed negotiations with opposition parties, the government announced in March that legislative elections would be held in May. Twenty-seven of the country's sixty-two officially registered parties contested the elections. A total of 64.8 percent of registered voters abstained from voting. Of the 107 seats of the Assemblée des Députés du Peuple (ADP), the ODP/MT won seventy-eight, the CNPP/PSD twelve, the RDA six, the ADF four, the PAI two, and five smaller parties one each. In June Compaoré appointed Youssouf Ouédraogo, a young economist, prime minister. Ouédraogo's government counted thirteen ODP/MT members and merged seven parties. Herman Yaméogo was appointed minister of state.

Social tensions resurfaced in late 1992 and 1993. Trade unions opposed the adoption of austerity measures in the context of Burkina's structural adjustment program, adopted in 1991. Students demonstrated against the reduction of their grants and for the payment of scholarship arrears. Amid social and labor unrest, the first ordinary session of the National Assembly opened on 9 April 1993 under the chairmanship of Arsène Yé Bognessan. It approved the establishment of a Chambre des Représentants, a 120-member, unelected consultative chamber renewable every three years, provided for by the constitution. In May it was announced that the CNPP/PSD had split up, with a faction led by Joseph ki-Zerbo founding the Parti pour la Démocratie et le Progrès (PDP). As a result, the CNPP's parliamentary representation shrunk from twelve to six seats, with the PDP taking the balance, further weakening the opposition to Compaoré. In September it was reported that Maurice Yaméogo had died.

Minor government reshuffles took place in September 1993 and January 1994, and on 20 March 1994 Roch Marc Christian Kaboré, the thirty-seven-year-old former chairman of the ODP/MT, was appointed prime minister in place of Youssouf Ouédraogo. This was seen as a political move by the president to please and further co-opt the ODP/MT into endorsing the rigors of the January devaluation of the CFA franc and of the World Bank–sponsored structural adjustment program. Two days later Kaboré appointed a twenty-three-member government, with Kanidoua Naboho heading defense, Herman Yaméogo responsible for "integration and African solidarity," Salif Diallo in charge of special duties of the presidency, Zéphirin Diabré in economy and finance, and Ablassé Ouédraogo in external relations.

## The Roots of Instability

Burkina's political instability is largely the result of a disjuncture between the state and civil society. Created by French colonialism and imposed as a structure of military occupation, political oppression, and economic exploitation, the state—appropriated by collaborating political elites upon independence—is historically illegitimate from a societal perspective.[62] In countries where the process of state formation has been endogenous (as in many European countries), states have evolved out of their relationship with civil societies, and the latter have often been sufficiently autonomous and articulate to help mold and legitimize the former. In Burkina as elsewhere in Africa, however, the state "has been deliberately set up *against* civil society rather than evolved in continual conflict with it."[63] As a consequence of society's challenge to their legitimacy, political elites in positions of state power have in many instances felt the need to neutralize civil society. In Burkina these attempts at state hegemony over society, whether confrontational or cooperative, have been only partly successful because of the state's institutional weakness and because of the continued strength of civil society. The failure to tame society has resulted in systematic changes in leadership. New leaders have tended to distance themselves from their predecessor's approach, creating what Otayek has called the "swing" between confrontation and cooperation that has characterized Burkina's political instability.[64] Generally, the agent of this swing has been the state institution in possession of physical force: the military.

The case for forceful hegemony was made by Maurice Yaméogo and Saye Zerbo's authoritarian regimes and by Thomas Sankara's attempts at totalitarianism. Theirs were instances of confrontation between state and society. The case for cooperative, or co-optative, hegemony was made by Sangoulé Lamizana, Jean-Baptiste Ouédraogo (briefly), and now Blaise Compaoré. Here, the state interacts with society through networks of clientelism under the formal setting of parliamentary democracy. In the latter situation society attempts to co-opt the state just as the state tries to co-opt society.

Yaméogo inherited the state from France—but not the French power to back it up. In the absence of historical legitimacy (which lay in institutions now belonging to civil society), absolutism and dictatorship increasingly provided the foundations of state power. Power-sharing became impossible, for it threatened the state with penetration by segments of civil society. Yaméogo outlawed political opposition, sterilized ethnic power, and attempted to absorb trade unions into his monolithic project. In short, his state "sought to remodel the political and social construction that it inherited so as to subject it to its order and substitute a national space of identification for particular ethnic or regional solidarities," and it sought "the elimination of any structure capable of constituting a way out from state domination."[65] The most vocal segments of society took to the streets and Yaméogo's project failed.

The subsequent regime of Lamizana marked the rejection of confrontation to the benefit of cooperative control, or what Otayek refers to as the "domestica[tion of] the institutions of civil society."[66] Lamizana allowed for the resumption of clientelist networks through which state and society communicated and interacted. Clientelism, rooted in the Africans' lack of allegiance for their state, is a system in which political support is given in exchange for parcels of power and redistribution of wealth outside the formal political institutions. It was the instrument of state-society cooperation under Lamizana as in general under any multiparty system in Burkina. The superficiality and rapid breakups of political alliances are indicative of the irrelevance of political programs and ideologies and of the importance of alliances and competition among different client networks. Witness Joseph Conombo's PSEMA's split from Ouézzin Coulibaly's RDA in 1958; the rivalry between Gérard Kango and Joseph Ouédraogo from 1970 to 1974, though they both belonged to the same party; the birth of the FPV in 1978, a politically odd alliance between Joseph ki-Zerbo and Joseph Ouédraogo; and Herman Yaméogo's repeated participation in Compaoré's governments in the 1990s.

Lamizana's brand of cooperative domination was sporadically interrupted by resurgences of confrontation when its paralyzing effects on the formal political institutions were too great. This occurred in 1970 with the announcement that the army would stay in power another four years and in 1975 with the aborted attempt at reinstalling a single party. Yet it took Saye Zerbo's coup in 1980 to put an end for good to Lamizana's benevolence and resolutely turn back to forceful hegemony. Like Yaméogo before him, Saye Zerbo tried to force Voltaic society to change and crushed expressions of pluralism by imposing censorship and banning strikes. Client networks fought back with the 1982 coup, which brought to power the CSP and, with it, Somé Yorian's goal to return to civilian rule.

This swing was short-lived, as Sankara prevented its full-fledged implementation and began the strongest confrontational campaign the Burkinabè state had ever waged against its society. The CDRs penetrated society down to the smallest villages and forcibly replaced ethnic structures in their local leadership and judicial functions. The TPRs broke down the client networks. Propaganda stigmatized nonrevolutionary classes and ideologies. Mobilization campaigns left little room for independent activities. Yet for all the commitment of its leaders, the CNR was defeated by society as it proved incapable of "imposing upon the peas-

TABLE 3.1  Abstention in Burkina's Electoral Consultations

| | Type of Poll | Abstention Rate (% of registered voters) |
|---|---|---|
| 1965 | Presidential | 1.7 |
| 1965 | Legislative | 2.6 |
| 1970 | Constitutional referendum | 24.1 |
| 1970 | Legislative | 51.8 |
| 1977 | Constitutional referendum | 28.6 |
| 1978 | Legislative | 61.7 |
| 1978 | Presidential (1) | 64.8 |
| 1978 | Presidential (2) | 56.5 |
| 1991 | Constitutional referendum | 51.0 |
| 1991 | Presidential | 74.7 |
| 1992 | Legislative | 64.8 |

SOURCES: Economist Intelligence Unit, *Country Report: Togo, Niger, Benin, Burkina*, various issues, 1988–1992; Pierre Englebert, "Burkina Faso: Recent History," in *Africa South of the Sahara 1995* (London: Europa Publications, 1994), 191–195; Pierre Englebert, *La Révolution burkinabè* (Paris: L'Harmattan, 1986), 36–57; Claudette Savonnet-Guyot, *Etat et sociétés au Burkina: Essai sur le politique africain* (Paris: Karthala, 1985), 149–172.

ants the disruption of their traditional hierarchies by the elimination of customary authority."[67]

Thus deprived of the social foundations it had counted on and cut off "from the ... machinery ... that established communication between state and society, the CNR condemned itself to being based exclusively on the minuscule and fragile political coalition that its accession had propelled to the summit of the state."[68] When the coalition broke down on obscure ideological grounds, the CNR and the revolution sank with it. Blaise Compaoré and the fourth republic have brought back an era of cooperation. Yet his regime has already been plagued by the problems of earlier republics: labor unrest, political deadlock, and the ODP/MT's hegemonic and confrontational tendencies. There is no indication at this point that the fourth republic will be more successful than its predecessors at reconciling state and society.

Why do both cooperative and confrontational approaches to state-society relations fail in Burkina? Cooperation, in its democratic form, fails because Burkina lacks one of its basic elements: consensus. Cleavages are stronger than consensual elements, causing social and political pluralism to be a force of division rather than cohesion.[69] The weakness of consensus derives from the lack of identification of society with the institutions of the state, whether democratic or not. The state cannot successfully confront society because of the Burkinabè's enduring opposition to the state's hegemonic aspirations. Civil society continuously challenges the state. Jean-François Bayart has identified several instruments of the social challenge of the state in Africa, among them strikes, abstention in elections, revolts, emigration, political tracts, humor, and the "sabotage of the instruments of political control."⁷⁰ All have been present in Burkina: Strikes have been numerous and have often influenced political outcomes; voter abstention in elections has been remarkably high and has increased over time, marking society's further challenge to the state at each phase of the latter's quest for legitimacy (see Table 3.1); revolts occurred in 1958 when the *mogho naba* failed to force the nascent state into a symbiosis with Mossi society and in 1966 when the Burkinabè took to the streets to find an outlet for their pluralism; and emigration, political tracts, and humor are pervasive features of Burkina's society and polity. As for the "sabotage of the instruments of political control," it was a widespread practice when the sons of village chiefs were elected to village CDRs under Sankara.

Yet while civil society is strong enough to challenge the state, it remains too weak to offer an alternative. It can initiate the swing but is unable to stop it. Colonization (and, earlier, slavery) have contributed to weakening civil society. Other debilitating factors are its ethnic and linguistic multiplicity, which play against consistent political attitudes and coherent actions, and its low level of literacy (see Chapter 5), which prevents it from ably playing the game of politics.

## Notes

1. Some of the arguments developed in the following sections appeared in my earlier book, Pierre Englebert, *La Révolution burkinabè* (Paris: L'Harmattan, 1986).

2. Philippe Lippens, *La République de Haute-Volta* (Paris: Berger-Levrault, 1972), 19.

3. Ibid., 20.

4. *Jeune Afrique Plus*, 8 (June 1984): 63–64.

5. Daniel Miles McFarland, *Historical Dictionary of Upper Volta* (Metuchen, N.J.: Scarecrow Press, 1978), 157.

6. René Otayek, "Burkina Faso: Between Feeble State and Total State, the Swing Continues," in Donal B. Cruise O'Brien, John Dunn, and Richard Rathbone, eds., *Contemporary West African States* (Cambridge: Cambridge University Press, 1989), 16.

7. Lippens, *La République*, 22.

8. McFarland, *Historical Dictionary*, 45.

9. François D. Bassolet, *Evolution de la Haute-Volta* (Ouagadougou: Imprimerie Nationale, 1968), 122.

10. *Le Monde*, 4 January 1966, 4.

11. Yaméogo was later arrested and condemned to five years' imprisonment in 1969. He was released in 1970 but remained deprived of civil and political rights until 1991. He died in 1993.

12. McFarland, *Historical Dictionary*, 123.

13. Quoted by Larba Yarga, "Le Tripartisme dans le droit public voltaïque," *Le Mois en Afrique*, 174–175 (June–July 1980): 120.

14. Quoted in ibid.

15. The RDA was in fact the Union Démocratique Voltaïque–Rassemblement Démocratique Africain (UDV-RDA). With the merging of the PRA and UNI, it became the PDV-RDA.

16. Larba Yarga, "La Fin de la troisième république voltaïque," *Le Mois en Afrique,* 182–183 (February–March 1981): 47.
17. See the section on the CNR in this chapter.
18. *Le Monde,* 27 November 1980, 3.
19. Ibid., 9 December 1980, 8.
20. Ibid., 12 December 1980, 7.
21. *Afrique Contemporaine,* 113 (January–February 1981): 13.
22. He was also a relative of Joseph ki-Zerbo.
23. Larba Yarga, "Les Prémices à l'avènement du Conseil National de la Révolution en Haute-Volta," *Le Mois en Afrique,* 213–214 (October–November 1983): 25.
24. In French the word *redressement,* which I translate as "recovery," has the additional connotation of redressing a tort, righting a wrong, streamlining looseness.
25. Claudette Savonnet-Guyot, *Etat et sociétés au Burkina Faso: Essai sur le politique africain* (Paris: Karthala, 1986), 176.
26. Press conference, 21 August 1983, quoted in Yarga, "Les Prémices," 32.
27. "Meeting de la Vérité," *Carrefour Africain,* 772 (2 April 1983): 10.
28. Ibid., 15.
29. Press conference of 28 June 1983, *Carrefour Africain,* 758 (1 July 1983): 10.
30. See on this latter point the article by a friend of Sankara, the late Mohamed Maiga, "Ouaga sur le qui-vive," *Afrique-Asie,* 301 (1 August 1983): 29.
31. ULC's full name was Union des Luttes Communistes Reconstruite (ULCR), a reference to an earlier split in the party's history. During a later split, in 1987, the departing faction took on the original name but became known as ULC–*La Flamme,* alluding to the title of its newsletter.
32. The Organisation Communiste Voltaïque was founded in 1971. See the book by the late Babou Paulin Bamouni, *Burkina Faso: Processus de la révolution* (Paris: L'Harmattan, 1986).
33. "O Captain, Their Captain," *Economist,* 305, 7521 (24 October 1987): 52–54.
34. Sennen Andriamirado, *Sankara le rebelle* (Paris: Jeune Afrique Livres, 1987), 12.
35. See Sankara's resignation letter in Bamouni, *Burkina Faso,* 173.
36. For a collection of his most representative speeches, see Thomas Sankara, *Thomas Sankara Speaks: The Burkina Faso Revolution, 1983–1987,* trans. Samatha Anderson (New York: Pathfinder, 1988).
37. René Otayek, "Rectification," *Politique Africaine,* 33 (1989): 3.
38. Conseil National de la Révolution, *Discours d'orientation politique (DOP)* (Ouagadougou: Ministère de l'Information, 1983).
39. See Charles Kabeya-Muase, *Syndicalisme et démocratie en Afrique noire: L'expérience du Burkina Faso, 1936–1988* (Paris: INADES/Karthala, 1989), 188.
40. See Babou Paulin Bamouni, *Principes d'action révolutionnaire* (Ouagadougou: Direction Générale de la Presse Ecrite, 1984), 14. The Burkinabè leaders were also very much influenced in their analysis by Amilcar Cabral; see his *Unity and Struggle: Speeches and Writings* (New York: Monthly Review Press, 1979), 134–137.
41. Pascal Labazée, "Discours et contrôle politique: Les avatars du sankarisme," *Politique Africaine,* 33 (1989): 11–26.
42. Labazée points out that in a later interview to the French periodical *La Tribune de l'Economie,* Sankara acknowledged that the working class did not exist in Burkina Faso; ibid., 15.

43. For a more complete analysis of the CNR's ideology, see Englebert, *La Révolution*, 97–119.

44. Conseil National de la Révolution, *DOP*, 28.

45. From the CDR propaganda organ *Lolowulen*, June 1986, 30–32.

46. Given the structural weaknesses of African states, full-fledged totalitarianism is most often beyond their leaders' means.

47. *Libération Afrique-Caraïbes-Pacifique*, "Spécial Haute-Volta," 22 (July–September 1984): 6.

48. The PCRV refused to recognize the country's new name, saying it fostered the illusion of independence in a neocolonial setting. The PCRV never changed its own name.

49. Otayek, "Burkina Faso," 19.

50. Ordonnance 84-2-CNR-PRES, *La Justice populaire au Burkina Faso* (Ouagadougou, Ministère de la Justice, 1985), 17.

51. Quotation of CDR secretary general Pierre Ouédraogo in *Le Monde*, weekly edition, 5–11 July 1984, 4.

52. For more details on the "clarification," see Englebert, *La Révolution*, 126–128, and Pascal Labazée, "La Voie étroite de la révolution au Burkina Faso," *Le Monde Diplomatique* (February 1985): 12–13.

53. See Valère D. Somé, *Thomas Sankara: L'Espoir assassiné* (Paris: L'Harmattan, 1990), 15.

54. None of these parties had large memberships but this did not bother them, as they thought of themselves as vanguard organizations. According to Somé, the UCB had only twenty-four members; see Somé, *Thomas Sankara*, 145.

55. August 1987 interview with a Cuban journalist, reprinted in Sankara, *Thomas Sankara Speaks*, 229.

56. Economist Intelligence Unit (EIU), *Country Report: Burkina Faso*, 4, 1987, 35; *Africa Contemporary Record*, vol. 20 (New York: Africana Publishing, 1989), B11.

57. *West Africa*, 29 February 1988, and Agence France Presse, *Bulletin quotidien d'Afrique*, 19 February 1988. See also United States Department of State, *Country Report on Human Rights Practices for 1987* (Washington, D.C.: Government Printing Office, 1988), 24.

58. Front Populaire, *Statuts et Programme d'Action* (Ouagadougou: Imprimerie Nationale, 1988).

59. *Le Monde*, 16 October 1988.

60. Radio Broadcast, *Radio Télévision du Burkina*, 20 September 1989.

61. *Marchés Tropicaux et Méditérranéens*, 9 March 1990, 680.

62. This dichotomy between state and civil society exists in most African countries. See Donald Rothchild and Naomi Chazan, eds., *The Precarious Balance: State and Society in Africa* (Boulder: Westview Press, 1988), and Jean-François Bayart, "Civil Society in Africa," in Patrick Chabal, ed., *Political Domination in Africa: Reflections on the Limits of Power* (Cambridge: Cambridge University Press, 1986), 109–125).

63. Bayart, "Civil Society," 112. See also C. S. Whitaker, "Doctrines of Development and Precepts of the State: The World Bank and the Fifth Iteration of the African Case," in Richard L. Sklar and C. S. Whitaker, eds., *African Politics and Problems in Development* (Boulder: Lynne Rienner Publishers, 1991), 333–336, and Basil Davidson, *The Black Man's Burden: Africa and the Curse of the Nation-State* (New York: Times Books, 1992), 10–12.

64. Otayek, "Burkina Faso," 13–30.
65. Ibid., 15–16.
66. Ibid., 17.
67. Ibid., 21.
68. Ibid., 26.
69. See Robert Dahl and Charles Lindblom, *Politics, Economics and Welfare* (New York: Harper and Brothers, 1953), 287–348.
70. Bayart, "Civil Society," 113.

# 4

# THE ECONOMY OF GROWTH AMID POVERTY

With a per capita gross national product (GNP) between $200 and $300, Burkina is among the poorest countries on earth.[1] It is also a small economy in absolute size, its GNP less than half of 1 percent of that of the United States in 1992. Relying mostly on the production of cereals for local consumption and exports of gold and cotton, it has few other resources than its work force, employed at about 90 percent in an agricultural sector highly dependent on erratic rainfalls. Yet despite its poverty and limited size, Burkina's economy is somewhat better balanced and has been growing on average faster than that of many of its wealthier neighbors. In the 1980s alone Burkina's real gross domestic product (GDP) grew at an average of 4.2 percent a year. Minus a 2.6 percent annual population growth rate, Burkina actually recorded annual per capita growth of 1.6 percent, a remarkable performance not only for West Africa but by overall sub-Saharan African standards. Although these statistics hide a bumpy business cycle with dramatic year-to-year fluctuations, data from the early 1990s suggest a continuation and possible improvement of this performance.

This chapter looks more closely at this apparent paradox of growth amid poverty. After placing Burkina's economy into historical and cultural perspective, it goes on with a descriptive profile of each sector and ends with a macroeconomic overview covering issues such as economic instability, monetary dependence, foreign aid, structural adjustment, and migration.

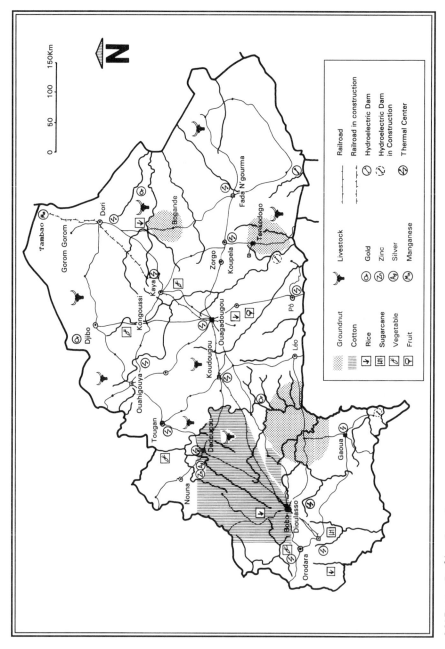

MAP 4.1  Burkina Faso: Economy

## The Colonial Economy

Because of its limited natural resources and harsh climate, Upper Volta was relegated to the back burner of French colonial economic development. Until about World War I, the French were merely a drain on the local economy. Wary of the potential budgetary consequences of its colonial advances, France had passed a law in 1900 requiring that colonial governments be supported by the colonized. For local administrators, this meant taxing the indigenous. Taxation was carried to such an extreme in Burkina that it turned into a systematic expropriation of the assets of Africans. Taxes were first collected in seeds, livestock, and cowries and later in French francs.[2] Louis Tauxier reviews the consequences of this level of taxation: "The Mossi were compelled to resort to trade in order to obtain the French money they did not have. ... To meet this difficult situation the Mossi now organize small caravans in the villages, and send the young men ... to sell cattle, sheep, goats, asses, horses, and bolts of cotton. They bring back with them either French money or kola nuts."[3] For lack of administrative resources and income data, the French resorted to a head tax, irrespective of wealth and income, and to an ad valorem tax on trading caravans that passed through the colony. By forcing people to sell their assets, the French tax burden destroyed the foundations of the economic system in which it operated while simultaneously providing the colonial administration with the means to establish a new economy. The French improved transport and communication infrastructures and introduced cotton around 1900. But with the introduction of cotton came the disruption of traditional cultures, compounding drought and in turn causing famines and epidemics in 1908 and 1914.[4] The French did little to help the Voltaics.

With the creation of the colony of Upper Volta, Governor Hessling undertook several economic development projects. Construction boomed in Ouagadougou to support the new administration: Roads, government buildings, schools, a hospital, and a football stadium were built in the new capital.[5] The commercial production of cotton was also encouraged. But the exploitation of Voltaic labor continued, forcing "thousands of Mossi and other Voltaics [to flee] Ouagadougou and the labor recruiters in the countryside and [migrate] to the Gold Coast" during the Hessling years.[6] Increases in taxation continued as well. The culture of rubber, unsuccessful earlier in the century, was resumed in the late 1920s, and groundnuts were introduced. In 1930, however, the worldwide recession reached Upper Volta, and prices for its commodities collapsed. A drought compounded the crisis in 1931. Deprived of many of their assets and weakened by emigration, Voltaics fared poorly in the slump that eventually turned into a full-fledged famine. The administration took stock of the economic failure of Upper Volta and the heavy emigration—which betrayed its comparative labor advantage—and partitioned the colony.

The partition of Upper Volta marked a return to exploitation at the expense of development. The main economic reason behind the partition was the French attempt to redirect the migration of workers away from the Gold Coast and toward its own colonies. Côte d'Ivoire benefited from Voltaic labor in agriculture (mostly in coffee and cocoa plantations and in the timber industry) and for its infrastructure projects (the port of Abidjan, the railway to Bobo-Dioulasso). Soudan put Voltaics to work on agriculture projects around the Niger River and on the railway between Bamako and Dakar. Although most of these projects had used Voltaic labor before 1933, the colony's partition increased the flow of workers and facilitated their administration by the French. Meanwhile, Upper Volta's development remained neglected and hampered by the depletion of its labor force.

In the period that followed World War II and the re-creation of Upper Volta, the French renewed their earlier emphasis on general infrastructure and agricultural development. By 1954 the railroad from Abidjan reached Ouagadougou. Emigration continued, however, as employment prospects in Côte d'Ivoire and Ghana still exceeded those in Upper Volta. Agricultural credit was made available, and irrigation works were begun (including the first studies for the future Sourou Valley scheme), but cash-crop production remained at low levels, with annual sales of no more than 4,000 tons of cotton seed.[7] Livestock production, in contrast, increased substantially, making Upper Volta the first West African producer of meat. The 1950s witnessed the beginning of the industrial exploitation of Poura's gold and the setting up of the Service des Mines. The first studies for the exploitation of Tambao's manganese were also undertaken.

Colonial economic policy had thus two contradictory effects on Upper Volta's economy. First, it acted as a drain, both in terms of taxation and exploitation of local assets and in terms of migration. Colonial-administered migration redistributed Upper Volta's labor assets to other colonies, such as Côte d'Ivoire, which were better endowed with natural resources, thereby increasing regional inequality among French colonies. Second, colonialism gave Upper Volta the foundations of a "modern" economy, with the introduction of cash crops as a source of foreign exchange and railway access to the ocean, of major importance with regard to the country's landlocked position. Most of all, colonialism transformed the local economy by turning peasants into workers, sedentary people into migrants, and rural dwellers into urban dwellers, fostering commercial exchange at the expense of subsistence and altering regional trade patterns. To further appraise the impact of colonialism on Voltaics and before looking at what Burkina did with its colonial economic legacy, this chapter now briefly turns to the economic culture of the peoples of Burkina.

## Culture and Economic Development

Although Burkina's successive leaderships have wanted to modernize the country through economic growth and development, there are cultures within the country that have not traditionally stressed the same objectives and have instead gen-

## The Economy of Growth amid Poverty

erated economic institutions based on different principles. A brief overview of some of these is of interest not only because they are part of Burkina's economy but also because cultural economic attitudes may provide clues to understanding overall economic performance.

In his book *Structures of Social Life,* Alan P. Fiske uses the Mossi as an example of societies where noneconomic motives dominate social relations.[8] Fiske illustrates how the Mossi (and some other African groups) interact mostly on the basis of "communal sharing" and "equality matching" rather than "market pricing" (by which Fiske refers to the rational self-interested utilitarianism that characterizes *Homo economicus* in Western literature).

According to Fiske, communal sharing is the paramount social motive behind the Mossi's economic behavior: "In their social relations, Moose only occasionally exhibit individualism, egoism, competition, maximizing intentions, or operate in a framework of free choice and contractually based obligations. . . . Moose prefer to share crucial resources."[9] This conclusion derives from the observation of how Mossi allocate their essential scarce resources: food, water, and land. Everyone shares in the production of food, grown primarily in collective fields, and in its consumption, irrespective of the individual's productivity. Only a small proportion of agricultural output is actually commercialized. The rest "goes into noneconomic activities whose goals involve communion with ancestors and other religious beings, food-sharing solidarity, gestures of support and sympathy, as well as fealty and obeisance."[10] As for water, no doubt one of Burkina's most precious resources, women often walk several kilometers a day and men risk their lives digging ever deeper wells to find it. And yet, according to Fiske, "this shortage usually does not lead to overt competition between individuals over water, or to any system of rationing. . . . For the most part, access to the village well is free to all comers, including outsiders."[11]

With respect to fertile land, high population pressure has added to its natural scarcity and made it ever more valuable. However, "anyone and everyone is welcome to move into a Moose village. . . . Land on which to build a compound . . . is always given for the asking. . . . Moose villagers never sell or rent land. Any land that is fallow is freely available to whoever wants to cultivate it."[12] In other words, owners of land do not make a profit from its scarcity. Whatever land they cannot exploit by themselves is available for anybody else to work without sharing the rewards—beyond a token gift to the owner. Other sources indicate that this may be true throughout Burkina, irrespective of ethnic affiliation.[13] Fiske concludes that the Mossi's "paramount goals concern mutual solidarity, a sense of common identity and belonging, unity, and kindness."[14]

In addition to sharing, the Mossi tend to structure their economic activities in ways that safeguard and reproduce equality among themselves. Fiske uses the example of agricultural work parties, where a premium seems to be placed on working together at the same pace ("the joy of laboring in synchrony") rather than individual efficiency.[15]

Yet there are three potential flaws in Fiske's analysis. The first one relates to the sheer poverty prevalent in Mossi society. Can market-pricing attitudes be adopted when the impossibility to acquire a resource could lead to death (from thirst or starvation)? Do not Western cultures, too, consider a subsistence level of essential resources as beyond market pricing (e.g., food stamps)? Another possible caveat is that the analysis is valid mainly at the village level, which is not so far from the kin level. Then the case could be made of the universality of non-market-pricing and of communal sharing within families. This qualification of the Mossi attitude Fiske describes—as valid only in the extended family setting—is supported by evidence that the Mossi act quite differently with the Peuls, for example, whose cattle they sometimes prevent from gaining access to their water.[16] Finally, several authors have demonstrated that solidarity and "gift-giving" are equivalent to insurance mechanisms and, as such, are rational ways to cope with risk in low-populated, semiarid regions of Africa. In addition, the absence of property rights over land is the normal consequence of the relative abundance of land and may be expected to change with population growth. On the other hand, these authors stress that because of the scarcity of people, such societies have developed property rights in humans in the form of slavery, which certainly cannot derive from the "unity and kindness" Fiske highlights.[17] Although these observations certainly qualify Fiske's findings, they do not fully invalidate them, for there do indeed appear to be differences in the pervasiveness of non-market-pricing attitudes in Mossi and some other Burkinabè and African societies versus more utilitarian societies.[18]

If one bears the above restrictions in mind, Fiske's observations may lead to some hypotheses on the low level of economic development of Mossi society. First, communal sharing appears antithetic to accumulation. Since "Moose are not trying to maximize production or make a profit" but "are simply trying to produce what they need to feed themselves and share with their kin and neighbors,"[19] they tend not to generate any surplus value or saving and thus fail to bring about capital accumulation and development. From this one can hypothesize that Mossi culture could just not be interested in economic development in its utility-maximizing, growth-oriented sense (see also Chapter 2, with regard to precolonial Mossi Kingdoms). It is indeed revealing that even those who are in a position to save through their increased access to the monetized economy, such as migrants to Côte d'Ivoire, virtually redistribute their earnings to the village, usually in the form of nondurable consumer goods (with a substantial part also paying for the village or family's head tax). Titinga Frédéric Pacere, a Mossi himself, makes the point that Mossi society bans individual accumulation "that could differentiate" one Mossi from another.[20] This is also supported by the absence of much visible economic stratification in Mossi villages. In addition, it appears that when village people do engage in market-pricing activities, they may not always set their prices and quantities in ways that maximize profit. Fiske again reports how several craftsmen and traders he interviewed had no idea whether they were making a profit or suffering a loss.[21]

Equality matching may represent another impediment to economic development, as it encourages rejection of technological progress. For example, there may be no incentive to adopt new agricultural technologies if they are an obstacle to communal work parties in the fields. But there are indications that Fiske's case is extreme and that many Mossi and members of other ethnic groups have been exposed to increased mechanization (from animal to motorized traction) and have assimilated different production techniques. It is true, though, that Burkina's agriculture is still hardly mechanized and functions at low levels of productivity.

As mentioned earlier, the Mossi are not alone in sharing a "nondevelopmental" economic culture. Similar features are found at least among the Gurmanche, the Bobo, and the Lobi. The Birifor, who are a branch of the Lobi, even seem to have gone a step further by having adopted a system that deliberately interrupts accumulation on any family farm (*yir*) before it becomes threatening to others.[22]

As in the political sphere, a disjuncture thus appears between Burkina's state, which operates in a *rational* economic setting and is officially interested in accumulation and growth, and its societies (even though market rationales are likely to have penetrated ethnic communities beyond what Fiske concedes). This is not so much a conflict as an inconsistency between the aspirations of state and civil society. It sheds a certain light on how official development policies may fail to meet the needs of local communities and how the latter, in turn, may fail to provide the foundations for state-sponsored development. As such, it contributes to the economic unsteadiness of Burkina's statehood.

## Agriculture as Livelihood and Constraint

### General Conditions

In 1960 agriculture accounted for about 62 percent of Burkina's GDP (see Table 4.1). About 90 percent of its work force was employed in the agricultural sector. In 1990 the numbers were respectively 32 percent and still about 90 percent, indicating the relative loss of productivity in agriculture. Since independence, and despite the progressive diversification of its output, Burkina has in fact remained an agricultural economy, providing a livelihood for the majority of the population but also constraining the country's development within the erratic limits of land and weather.

Indeed, the performance of the agricultural sector had been mostly disappointing until the mid-1980s. It began the 1960s with low productivity, little integration with other sectors, and few modern inputs. Most of the cultivable land was fallow or uncultivated, and the system functioned in near autarky, with on-farm consumption the rule and farmers purchasing few agricultural products.[23] Although production grew over time, its average annual increase of 2.2 percent from 1965 to 1986 remained below population growth. The situation changed in the 1980s. According to the World Bank, the agricultural sector grew by 4.3 per-

TABLE 4.1  Industrial Origin of Gross Domestic Product (as % of GDP)

|  | 1960 | 1965 | 1970 | 1975 | 1980 | 1985 | 1990 |
|---|---|---|---|---|---|---|---|
| Agriculture, livestock, etc. | 62 | 53 | 42 | 41 | 45 | 45 | 32 |
| Industry | 14 | 20 | 21 | 21 | 22 | 22 | 24 |
| Manufacture | 8 | 12 | 14 | 14 | 11 | 11 | 14 |
| Services | 24 | 27 | 37 | 39 | 32 | 33 | 44 |
| GDP at market prices | 100 | 100 | 100 | 100 | 100 | 100 | 100 |

SOURCES: World Bank, *Burkina Faso: Economic Memorandum* (Washington, D.C.: World Bank, 1989); Banque Centrale des Etats de l'Afrique de l'Ouest, "Burkina: Statistiques économiques et monétaires," *Notes d'Information et Statistiques,* various issues; Institut National de la Statistique et de la Démographie, *Annuaire statistique du Burkina Faso, 1989–1990* (Ouagadougou: Direction des Statistiques Générales, 1991).

cent between 1978 and 1982 and 5.4 percent between 1982 and 1987—faster than GDP—mostly because of growth in cotton production. Furthermore, the late 1980s and early 1990s witnessed a substantial boom in cereal production. Throughout the 1980s cultivated area expanded and productivity finally took off, partly because of increases in fertilizer consumption and mechanization.[24]

Yet recent improvements do not make up for the structural vulnerability of Burkina's agricultural sector. The main constraints remain the shortage and irregularity of rainfalls, which range from 350 mm per year in northern Sahel to 1,000 mm in the southwest, and population pressure on land. The level of rainfall is of paramount importance to Burkina's agriculture, for the country's geography, with few permanent watercourses, severely limits the potential for irrigation (only 0.1 percent of agricultural land is under irrigation). Most crops are thus rain-fed. But the rainy season is short, lasting at best from May or June to October, and rains are often late and occasionally interrupted during the season. Since independence, several droughts, the worst in 1973–1974 and 1984–1985, have affected the country. As a result, output varies widely from year to year. In years of good rains (such as 1985, 1986, 1989, 1992, and 1994), Burkina easily produces a cereal surplus, but should the rains be delayed or insufficient (as in 1988, 1990, and 1991), it fails to reach even subsistence levels.[25] Cereal output shows a correlation with rainfall of $r = 0.71$ for the 1980s. Because of the importance of agriculture in GDP, a correlation coefficient of $r = 0.59$ is also observable between rainfall and GDP growth for the same period.[26] In other words, almost three-quarters of the yearly variation in agricultural production and more than half that in GDP are associated with fluctuations in the volume of rainfall. Very little mediates this relationship, which highlights the substantial vulnerability of the agricultural sector and of the whole economy. Since the 1980s, however, successive governments and nongovernmental organizations have taken measures to reduce this dependency on rain. The Sankara and Compaoré regimes have encouraged local low-cost water retention schemes and have pushed forward the Sourou Valley Rural Integrated Development Program, an irrigation scheme on a

The Economy of Growth amid Poverty 85

*Songhai granary in Ye for the storage of millet and sorghum, near Dori. Courtesy of the Centre National de la Recherche Scientifique et Technique.*

tributary of the Black Volta. Upon completion, the Sourou project should provide 40,000 ha of irrigated land.[27]

Humans are the second largest obstacle to the development of Burkina's agriculture. Population pressure in the central plateau upsets an economy formerly based on unrestricted land availability. It limits the area available for cultivation, shortens the time dedicated to fallow, and reduces the quality of the land. More recently, the economic expansion of the more fertile western and southwestern regions has induced additional internal migration, which threatens to push population problems to the west. This trend has been stimulated by the apparent eradication in the Volta basins of onchocerciasis (also called river blindness, a disease carried by flies that eventually blinds its victims) freeing up fertile land for resettlement since the late 1980s.

Food crops, cash crops, and livestock are the three main components of agricultural production. Food crops, mainly traditional cereals (such as millet, sorghum, and fonio), maize, and rice, occupy about 80 percent of the cultivated area. Cotton and groundnuts provide the bulk of cash crops. The contribution of *karité* (shea nut) to the revenue of cash crops is very uneven because of its three-year output cycle. Fruits and vegetables serve as both food and cash crops, and

TABLE 4.2  Cereal Output and Commercial Production of Main Cash Crops

| Crop year beginning in: | 1980 | 1981 | 1982 | 1983 | 1984 | 1985 | 1986 | 1987 | 1988 | 1989 | 1990 | 1991 |
|---|---|---|---|---|---|---|---|---|---|---|---|---|
| Food crops (thousands of tons) | | | | | | | | | | | | |
| Millet and sorghum | 898 | 1,102 | 1,050 | 1,003 | 1,011 | 1,417 | 1,690 | 1,480 | 1,826 | 1,640 | 1,514 | 1,870 |
| Millet | 351 | 443 | 441 | 392 | 417 | 631 | 679 | 632 | 817 | 649 | — | — |
| Sorghum | 547 | 659 | 609 | 611 | 594 | 786 | 1,011 | 848 | 1,009 | 991 | — | — |
| Maize | 98 | 132 | 111 | 72 | 78 | 142 | 158 | 131 | 227 | 257 | 217 | 296 |
| Rice (paddy) | 29 | 29 | 27 | 19 | 41 | 21 | 27 | 22 | 39 | 42 | 43 | 50 |
| Total | 1,025 | 1,262 | 1,188 | 1,094 | 1,130 | 1,580 | 1,875 | 1,632 | 2,092 | 1,939 | 1,774 | 2,216 |
| % change | 11.2 | 23.2 | −5.9 | −7.9 | 3.2 | 39.9 | 18.7 | −12.9 | 28.2 | −7.3 | −8.5 | 24.9 |
| GDP % growth (calendar year) | 1.7 | 4.4 | 2.2 | −1.2 | −2.1 | 9.8 | 15.7 | 1.2 | 6.1 | −0.4 | 1.3 | 4.8 |
| Cash crops (tons) | | | | | | | | | | | | |
| Groundnuts | | 438 | 1,366 | 49 | 164 | 5,369 | 803 | 379 | 1,374 | 2,254 | 1,405 | — |
| Cotton (seed) | 62,539 | 57,534 | 75,287 | 79,287 | 88,134 | 115,491 | 169,593 | 148,015 | 145,898 | 152,000 | 189,543 | 190,000 |
| Cotton (fiber) | | 21,628 | 28,817 | 30,074 | 34,382 | 45,979 | 65,971 | 58,454 | 62,000 | 66,500 | 73,920 | 69,200 |
| Karité (shea nut) | 44,645 | 23,150 | 23,564 | 66,675 | 1,646 | 70,036 | 8,377 | 1,825 | 4,241 | 21,523 | 10,082 | 6,400 |
| Sesame | 6,013 | 8,017 | 5,675 | 4,591 | 8,670 | 4,571 | 4,353 | 656 | 1,383 | 47 | 1,428 | — |
| Sugar cane (thousands of tons) | | 290 | 298 | 275 | 286 | 275 | 275 | 272 | 272 | 277 | 273 | — |

SOURCES: Institut National de la Statistique et de la Démographie, *Annuaire statistique du Burkina Faso, 1989–1990* (Ouagadougou: Direction des Statistique Générales, 1991); Banque Centrale des Etats de l'Afrique de l'Ouest, "Burkina: Statistiques économiques et monétaires," *Notes d'Information et Statistiques*, various issues; Jacques Lecaillon and Christian Morrisson, *Economic Policies and Agricultural Performance: The Case of Burkina Faso* (Paris: Development Centre of the OECD, 1985).

their importance has been growing. Livestock, used primarily for exports and as a store of wealth, consists of nearly 3 million cattle, 9.5 million sheep and goats, and more than 22 million poultry.[28]

## Food Crops: The Ongoing Battle for Self-Sufficiency

Burkina's main food crops are rain-fed millet and sorghum, whose short cycles and resistance to drought are well adapted to the region's brief and erratic rainy seasons. These two cereals are used in the preparation of *to*—a puree that, accompanied by vegetable sauce and occasional meat or fish, is the basis of most meals—and *dolo*, the traditional sorghum beer. Maize, fruit, and vegetables provide households with important additional sources of nutrients. Most rice is imported, but local production is expected to increase with the development of the Sourou Valley project. Bread (the French *baguette*) is popular in urban areas, but wheat is not locally grown.

Output of millet and sorghum has grown steadily since before independence, despite competition for land from cash crops, which has restricted the area given over to millet and sorghum and shortened the length of fallow, especially in the Mossi plateau. Production averaged 560,000 tons in 1948–1957, 800,000 tons in 1961–1964, and 1 million tons in 1977–1981.[29] From 1981 to 1991 the average annual output was 1.418 million tons, with a substantial upward trend from 1985 onward (see Table 4.2). The steady upward trend in output continued in the early 1990s to reach more than 2 million tons. Recent productivity increases are in marked contrast to the first two decades of independence, when productivity was estimated to have improved by only 25 percent overall.[30]

The production of rice has not experienced the same record of growth. Average annual production amounted to 17,000 tons in 1948–1957 and 33,000 tons both in 1961–1964 and 1977–1981. From 1960 to 1980 the yield per hectare increased by only about 15 percent.[31] Furthermore, unlike millet and sorghum, rice stagnated in the 1980s, with an average production slightly below 33,000 tons for the period 1981–1991, forcing Burkina to remain an importer of Asian rice. Output growth seemed, however, to have resumed at the end of the decade: The average production for 1988–1991 amounted to 43,500 tons.

Maize is grown primarily by households on land that surrounds their compounds and is usually considered an adjuvant cereal. Its production, too, has grown steadily since the early 1980s, climbing from 98,000 tons in 1980 to 296,000 tons in 1991. Yams, sweet potatoes, cassava, and other fruits and vegetables are usually grown on small individual plots, and production estimates are unreliable and hard to come by.

Despite the progress and modernization of the last decade, Burkina's cereal sector remains inward-looking, subsistence-based, and technologically backward. Only a small fraction of output is marketed, and there is little exchange with other sectors of the economy, both in terms of output sales and input purchases.[32] There even seems to be little exchange within villages, for most compounds consume what they themselves have produced. Jacqueline Sherman showed that in the Mossi region of Manga, "a relatively market-oriented area, families sell an average of only about 11 percent of the sorghum, millet and rice they grow [and] buy grain equivalent to about 6 percent of what they produce."[33] In addition, technological development, which has affected the cash-crop sector, has bypassed most cereal farmers who still use the *daba*, the traditional hoe, as their only tool. This lack of market integration and technological progress increases the vulnerability of cereal farmers, who end up excessively dependent on their own local conditions and crops.

The overall climatic deterioration from independence to the mid-1980s no doubt accounted in part for the relatively poor performance of agriculture over the same period. A study by the Organization for Economic Cooperation and Development (OECD) reckons that a national rainfall index equal to 111 on average between 1960 and 1965 had plunged to 91 between 1975 and 1980, and a World Bank index of per capita food production fell from 100 in 1965–1967 to 84 in 1975, dragging the daily caloric intake per head down to 93 percent of need at the end of the 1970s.[34] The situation stabilized and actually improved following the 1983–1984 drought. Given available data, I estimate that the rainfall index averaged 93 between 1980 and 1989, and the World Bank contends that the food production index improved by 20 percent since 1980.[35] This comes as no surprise in view of recent output figures. Total cereal production reached 2 million tons for the first time in 1988, more than twice its 1980 volume. In the 1991–1992 campaign a record harvest of 2.2 million tons was reached, and in 1993–1994 a new peak was again achieved at 2.5 million tons.

Nevertheless, in good as in bad years, the overall food production picture hides substantial regional differences. The regions of Yatenga, Ouagadougou, and the north are usually in deficit, while the west and southwest experience frequent surpluses, and the east and southeast hover around self-sufficiency. Unfortunately, poor infrastructure often prevents the redistribution of excess output; as a result, the government has to resort to food aid and food imports.

Food aid, most of it in the form of cereals, increased from around 28,000 tons per year (2.5 percent of domestic production) in the mid-1970s to 73,000 tons a year (4.6 percent of domestic production) in the mid-1980s before contracting to 44,000 tons (2 percent of domestic production) in the early 1990s.[36] The high level of aid of the mid-1980s, in the wake of the drought of 1983–1984, illustrates how food inflows tend to continue even during surplus years, as donor reaction typically lags behind the actual drought situation. Food imports follow the same pattern: They increased by 37 percent, from CFAFr 32,051 in 1984 to CFAFr 43,921 in 1985, although the latter year witnessed a good harvest.[37] Food aid is either freely distributed (with the potentially perverse consequence of depressing the local market prices and acting as a disincentive for farmers) or sold at subsidized or full-market prices.

Faced with uncertain weather and crops, Burkina's farmers have developed strategies of income diversification and risk prevention. Thomas Reardon, Peter

Matlon, and Christopher Delgado have shown how rural people generate "purchasing power in non-cropping occupations,"[38] later using that revenue to purchase food, especially during the season immediately preceding harvest, when shortages are most severe. Reardon and his colleagues found the following distribution of sources of income for Sahelian and Sudanian households, repectively: agriculture 23 percent and 56 percent; livestock 22 percent and 6.3 percent; local nonfarming income (crafts, services such as *dolo* brewing, and commerce) 25 percent and 14 percent; nonlocal, nonfarm income (including temporary migrants' remittances) 22.6 percent and 16.8 percent; transfers from abroad 7.8 percent and 7.6 percent (of which 4.2 percent and 0.1 percent in food aid and 1.9 percent and 6.8 percent in remittances from permanent migrants).[39] In other words, there is more than farming and herding to Burkina's farmers and herders.[40]

## Cash Crops: The Tricks of World Markets

Burkina's main cash crops are cotton, groundnuts, *karité* (shea nut), sesame, and sugar. Livestock has also historically been a major export, and a share of the vegetable and fruit production is exported each year.

Cotton is the chief export crop, accounting for about 50 percent of overall export revenue and covering about 165,000 ha as of the mid-1990s. From a 3,000-ton annual production at independence, output grew to 60,000 tons a year in the early 1980s and further surged to reach almost 200,000 tons in the early 1990s.[41] This performance was due to the extension of the area under cultivation, the introduction of better seeds and fertilizers, and the stability of rains in the southwest region, the main cotton-producing region. Yet because of limited ginning capacity, lint output has remained at between 60,000 and 70,000 tons.

About one-third of Burkina's farmers are involved in the production of cotton. These small planters sell their output, at a fixed price announced before the campaign, to the partly state-owned company SOFITEX (Société des Fibres Textiles), which processes and markets it and also provides them with needed inputs. The harvest is essentially purchased by the French Compagnie Française pour le Développement des Fibres Textiles (CFDT), which owns part of SOFITEX.

Unfortunately for planters and for the state, revenue has not kept up with increases in output because cotton's highly volatile world price collapsed in the 1980s just as production was soaring. In dollars the international price of cotton went from an index of 110 in 1960 down to 100 in 1970 and then skyrocketed to 324 in 1980, only to disintegrate in the 1980s and 1990s, reaching a low of 186 in 1986 and standing at 211 in 1993. In CFA francs the collapse came later, as revenues were supported by the strength of the dollar in the first half of the 1980s. Yet if these prices are deflated by the consumer price index (CPI) and thus expressed in constant CFA francs, the index fell from 138 in 1960 to 100 in 1970, 83 in 1980, and 66 in 1990. As of 1993 it stood at a dismal 50.[42] Producer prices, however, rose from CFAFr 80/kg to CFAFr 112/kg for the 1993–1994 campaign as a result of the devaluation of the CFA franc. With the exceptionally heavy rains of 1994, an increased cotton output should combine with the devaluation to trigger record revenues in 1995.

Groundnuts have progressively lost importance as a cash crop. Overall production declined in the 1970s, mostly as a result of unfavorable producer prices. The recovery of producer prices in the 1980s triggered a relative surge in production in the second half of the decade, but groundnuts still accounted for a mere 0.5 percent of total exports in 1989. A substantial part of the harvest is now consumed as a food crop.

*Karité*, a nut used to make oil and butter, follows a natural three-year cycle, which makes it unreliable as an earner of foreign exchange (it accounted for 11.5 percent of export value in 1984 versus 0.37 percent in 1989). Output can vary tremendously, as it did between 1983 and 1985, when it fell from 66,675 tons to 1,646 tons before climbing back to 70,036 tons. In addition, its world price did not fare well in the 1980s.

Other contributors to foreign exchange earnings are sesame (from 0.5 percent to 4 percent of export revenue) and vegetables (around 2 percent). Sugarcane sustains the SOSUCO sugar refinery at Banfora and makes Burkina self-sufficient in sugar. Sugar is not exported, however.

Although cash crops remain the backbone of Burkina's external trade, they have also contributed to the instability of its economy. First, the country's lack of control of commodity prices has rendered it hostage to world market conditions, which have triggered substantial volatility in revenue from year to year (cotton revenue, for example, was CFAFr 19 billion in 1984, 11 billion in 1986, 20 billion in 1987, and 14 billion in 1989). Second, the expansion of cash crops has often happened at the expense of food crops. Good rains in recent years have hidden the potential problem triggered by the transfer of land formerly dedicated to cereal to the production of cotton. All agricultural exports combined rarely cover the cost of imports of foodstuff. Nevertheless, they provide the state with most of its foreign exchange earnings and a large part of its tax revenue.

## *Evolving Patterns of Livestock Production*

Next to cotton, livestock is the second most important agropastoral commodity. It was once Burkina's major export but currently comes in third place behind cotton and gold. Yet exports of live animals still account for between 4 percent and 10 percent of all export revenue. In addition, hides and skins contribute about 5 percent a year to total exports. The dwindling importance of livestock as an export has resulted from several factors, including the surge in cotton and gold output, the effects of drought on animal survival rates (it takes two years to replenish a herd after a drought), and the progressive sedentarization and transformation of the formerly nomadic Peul herders who traditionally managed livestock production. Recessions in the regional markets (such as Côte d'Ivoire and Ghana, where virtually all livestock is exported on hoof) and competition from Latin American (mainly Argentinian) meat and subsidized European meat have also hurt Burkina's livestock industry.

Apart from being an export commodity, livestock also serves as a form of saving and investment, as it grows and reproduces, weather permitting. It is rarely eaten but rather sold in poor years in order to purchase cereal.[43]

The traditional arrangement for cattle herding has been for sedentary farmers such as the Mossi, Bissa, and Gurmanche to entrust their cattle to the seminomadic Peul herders.[44] In exchange, all the milk produced by the cows becomes the property of the Peuls, who also receive small annual gifts and cash payments when they handle the sale of an animal. The advantages to both the Mossi and the Peuls are numerous.[45] For both, there are the benefits of specialization (the Peuls are experienced herders and the Mossi able agriculturalists whose labor force is better used farming). Contracting-out herding also allows the Mossi to keep the number of cattle they own secret "from both neighbors and government tax collectors."[46] Finally, the Mossi usually arrange for the herds to be pastured in their fields after the harvest, obtaining thereby natural fertilizer. For the Peuls, the first advantage is ownership of the milk, which they use as a source of food and as a source of income (Peul women sell milk and milk products such as yogurt on local markets). By paying the Peuls in milk, the Mossi are guaranteed against misuse of their cattle; the care of the livestock becomes as important for the herders as it is for the owners. The addition of the cattle from the Mossi to the Peul herds induces decreasing marginal costs and thus represents little extra effort. The Peul also benefit from the annual gifts (mostly food and clothing), and the cash payments compensate for the additional work of selling a Mossi's cow.

Observers of Burkina's rural life, however, have noticed that this pattern of herding has been changing since the 1970s. Today the Mossi are more likely to let their own children herd their cattle or to entrust it to other Mossi farmers who have larger herds and more experience. The Mossi selection of their own family members and people from the same ethnic group suggests that the Mossi have felt the need for additional monitoring. This evolving pattern coincides with an increased distrust between Mossi and Peuls. Mossi now complain about "the destruction of crops by cattle who accidentally enter into fields or gardens."[47] Although it is legally the responsibility of the herder to pay for the damage, "in fact, owners of cattle involved are expected to, and do make significant contributions to such settlements."[48] This was not a problem until population pressure on the Mossi plateau reduced the area available for grazing and transhumance and increasingly sedentarized the Peuls, making the traditional herding contract less attractive for the Mossi. If the Peuls are going to remain around the village all year, the Mossi may as well have their children take care of the cattle. Since the children share the family's interest in keeping the crops safe, they are likelier to supervise the animals more carefully, limiting crop destruction.

Another reason the practice of Peul herding has changed may be related to the increase in droughts since the early 1970s and the resulting need for the Peuls to diversify their activities. In this case the risk motive (income diversification) counteracts the transaction-cost motive (the monitoring of the Peuls, etc.). As a result, the Peuls have also lost some of their interest in proper herding, and the Mossi may have legitimate complaints about asset misuse and lack of care.

As further population pressure continues, it is unlikely that the Mossi will revert to the traditional herding arrangements. A more plausible outcome will be for herding to remain an intrafamily activity in areas of traditional farming (such as the central plateau and the east) and to become a wage contract activity in areas of more developed and industrialized agriculture (such as the southwest).

## Agricultural Methods and State Strategies

In the 1990s, as in precolonial times, small units dominate Burkina's agriculture. Some 600,000 family holdings, most of which do not exceed 5 ha and three-quarters of which are committed to millet and sorghum, provide 95 percent of the country's agricultural output; the rest is accounted for by state and private farms.[49]

The customary system of land tenure makes a distinction between ownership and use of the land. Property rights are usually held by and transferred within families, but fallow land is generally open for anyone to cultivate, irrespective of ownership. Occasionally, land may be transferred because the original owner finds it difficult to establish title over the land after another household has farmed it for a prolonged period.

The Mossi and most other farming ethnic groups work their fields communally, usually as a family unit, and share the production in a similar manner. Individuals also cultivate their separate fields when time permits, and women have their own plots of vegetables. There has been only limited technological change in the typical village since precolonial times. Many peasants still depend on the *daba* as their only tool, and inorganic fertilizer is used on only about 3 percent of the cultivated land.[50] In some areas, however, these customary patterns are changing. An OECD study has established that

> in the more advanced regions [the west and the areas surrounding the Black Volta] the traditional system is being called into question because of the progress of mechanisation.... Land then becomes a rare good with a market value, subject to individual ownership. Differentiation at the social level then makes its appearance along with the differences in the size of the holdings and the incomes obtained from them.[51]

In addition, although households and villages (including village cooperatives) still represent the bulk of Burkina's agriculture, another major change distinguishes today's agriculture from that of precolonial times: the intervention of the state, either as an associate of villagers in production and distribution; as a powerful seller, purchaser, and price-setter; or as an agricultural policymaker.

The state's first level of intervention is through public investments. The largest one, which has been on the agenda of many governments since the 1960s, is the Rural Integrated Development Program in the valley of the Sourou River, a tributary of the Black Volta, in the provinces of Sourou, Yatenga, and Passoré. The first phase of the Sourou project, a 714-km canal constructed entirely by local labor under the Sankara government (a period of extreme state intervention in agriculture and economic life in general), brings water from the Mouhoun to the Sourou reservoir also constructed by local labor. In the ongoing second phase, the government intends to establish around 40,000 ha of irrigated land, 28,900 ha of which will be in the Sourou Valley and 10,810 ha in the upper valley of the Mouhoun.[52] In general, between 10 and 25 percent of (mostly aid-financed) public investments go into agriculture.

Extension agencies represent another means of state action. The Organismes Régionaux de Développement (ORDs), founded in 1965, long fulfilled these functions. Dependent upon the Ministry of Rural Development, the eleven ORD centers provided services such as training, agricultural technique development, and input distribution. Suffering from poor management and inadequate staffing, they were dissolved in March 1987 only to be replaced two months later by twelve Centres Regionaux de Production Agro-pastorale.

The most pervasive state role in agriculture, however, is as seller of inputs and purchaser of harvests. That role is played by SOFITEX with regard to cotton production. Until the early 1990s the parastatal Office National des Céréales (OFNACER) was responsible for maintaining cereal stabilization and security stocks, but it had been running cumulative losses totaling several billion CFA francs and was liquidated as part of the structural adjustment program. Its action was complemented by that of the marketing board Caisse de Stabilisation des Prix des Produits Agricoles, which still exists. Not only have these parastatals run huge deficits over time, they have also generally failed in fulfilling their mission. Private traders currently distribute about 80 percent of the marketed production of cereal,[53] and price speculation is widespread in the months that precede a new harvest.

Finally, the state influences agriculture as a policymaker. Its agricultural strategy is a political issue and has changed with regimes and with their perception of the relative importance of rural versus urban dwellers. One instrument of agricultural policy is the setting of producer prices. Figure 4.1 illustrates the relationship among the consumer price index, salaries of urban dwellers (shown here through the minimum guaranteed salary, or SMIG), and producer prices for cereal and cotton since 1972. The graph clearly shows that cotton producer prices have generally lagged behind inflation, while salaries generally stayed ahead of it. As for cereal producer prices, they climbed well ahead of inflation until 1986, when they collapsed and entered a more volatile pattern. The graph shows the relative urban bias of all regimes. There was a trend in favor of agricultural producers under the CSP and in the first half of the CNR rule: Real salaries in the civil service fell, while cereal and cotton producer prices rose relative to inflation. The trend was suddenly reversed in 1986 as political problems (partly due to the loss of support from urban salaried groups) and fiscal pressure mounted against

FIGURE 4.1  Wages and Prices in Burkina

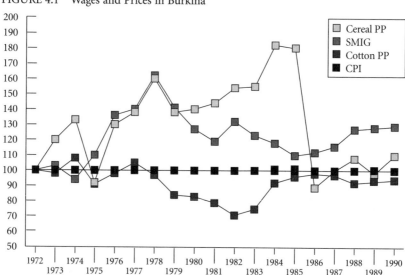

SOURCES: Economist Intelligence Unit, *Country Profile: Burkina, Togo* (London: EIU, 1994); International Monetary Fund, *International Financial Statistics,* various issues; Banque Centrale des Etats de l'Afrique de l'Ouest, "Burkina: Statistiques économiques et monétaires," *Notes d'Information et Statistiques,* various issues; Jacques Lecaillon and Christian Morrisson, *Economic Policies and Agricultural Performance: The Case of Burkina Faso* (Paris: Development Centre of the OECD, 1985); World Bank, *Burkina Faso: Economic Memorandum* (Washington, D.C.: World Bank, 1989).

the Sankara administration. The subsequent return of the urban bias has been further amplified under the FP and the fourth republic.

Policies such as setting producer prices affect not only the relative income of rural and urban dwellers but also act as incentives or disincentives for the volume of output. Producer prices are announced by the government ahead of the agricultural season and affect planting decisions by growers (see Table 4.3). Yet the relationship between producer prices and output (the "supply" response to prices) is more readily apparent with cash crops such as cotton than with cereals, where a large part of the output is not commercialized and thus escapes monetary incentives.

The policies of the CNR vis-à-vis the rural world were thus ambiguous. Even during the first years of his administration, when he meant to favor peasants, Sankara adopted policies that hurt their interests, such as the 1985 Agrarian and Land Reform Law, which nationalized land owned either as private property or according to custom. The reform officially purported to provide food self-sufficiency, dwellings for all, and a rational use of land from the point of view of productivity and social justice. Yet it questioned the customary management of land

TABLE 4.3 Evolution of Agricultural Producer Prices by Political Regime (in CFAFr/kg)

|  | Millet and sorghum | Rice | Cotton | Groundnuts | Sesame | Karité |
| --- | --- | --- | --- | --- | --- | --- |
| 1960–65 | 13 | 19 | 33 | 27 | 23 | 7 |
| 1966–79 | 20 | 32 | 38 | 34 | 37 | 13 |
| 1980–82 | 52 | 66 | 60 | 117 | 81 | 39 |
| 1983–87 | 63 | – | 91 | 121 | 86 | 48 |
| 1988–91 | 56 | – | 95 | 100 | 98 | 28 |

SOURCES: Jacques Lecaillon and Christian Morrisson, *Economic Policies and Agricultural Performance: The Case of Burkina Faso* (Paris: Development Centre of the OECD, 1985); Economist Intelligence Unit, *Country Profile: Burkina, Togo* (London: EIU, 1994); Banque Centrale des Etats de l'Afrique de l'Ouest, "Burkina: Statistiques économiques et monétaires," *Notes d'Information et Statistiques*, various issues; World Bank, *Burkina Faso: Economic Memorandum* (Washington, D.C.: World Bank, 1989).

and was apparently viewed with much hostility by most of the rural world—although the law actually never seems to have been implemented. As Bernard Tallet mentions, the reform was political and based on the CNR's beliefs in a new relationship between the state and the peasantry rather than technically and agronomically articulated. Tallet also demonstrates how the CNR in other respects did much less than it claimed to challenge the traditional relation between the state and the peasantry. The relative amount of resources the CNR allocated to the fertile and already privileged southwest region of the Black Volta did not significantly change from that of previous regimes, and old projects such as the Sourou were maintained.[54] One action the CNR did take in favor of the peasants was the repeal of the *impôt de capitation*, a poll tax on peasants.

Overall, the role of the state in Burkina's agriculture since independence was summed up by the World Bank in 1989 as "a complex system of institutions, regulations and mechanisms which give the Government a dominant role," involving "four key ministries, six marketing and stabilization boards, more than 20 decentralized agencies, and numerous parastatal enterprises for production, extension and research. Fiscal, price and trade regulations are both numerous and cumbersome."[55] Things began to change, however, with the adoption of an agricultural adjustment program with support from the World Bank in June 1992. The program aimed to liberalize the pricing system and the marketing of inputs, formalize the land tenure system, improve the management of food aid, and promote private initiative. The program is beginning to bring about far-reaching changes in Burkina's agriculture and in the role of the state therein. Above all it has provided for the privatization and the liquidation of several state companies and parastatals that dealt directly or indirectly with agriculture. In July 1992 the World Bank approved a loan for a "food security and nutrition project" with cofinancing from Germany and the UN World Food Program. In 1991 the government had already introduced a national "environmental management project" with another credit from the World Bank's IDA. The project sought to use a "participatory approach" to "stop and reverse the process of natural resource degradation in order to meet the conditions essential for sustainable agricultural production."[56] This project was timely, for the degradation of Burkina's environment accelerated in the 1980s. Human pressure has reduced the quality of the soil and water and the extent of the plant cover. The need for firewood and charcoal has led to the "exploitation of wood products in excess of the natural regeneration rate and consequent deforestation at the rate of 1,000 km² a year, with all the attendant effects on soil erosion and loss of surface water."[57]

## Mining and Industrial Underdevelopment

### Mining: A Fragile New Hope

Burkina's known mineral resources include gold, zinc, manganese, limestone, phosphate, and diamonds. Rumors of an oil deposit in the Agacher strip, which lies on the border between Burkina and Mali, were in part responsible for two short wars between the two neighbors in 1974 and 1985 but have yet to be substantiated.

Gold mining is virtually as old as humanity in Burkina. It was one of the region's main products of exchange from the eleventh to the nineteenth century, and gold panning has been practiced in the Black Volta Valley for centuries.[58] Gold mining became important to the modern economy in 1984, when the Sankara government decided to reopen the Poura gold mine. Located 60 km southwest of Ouagadougou, the mine had 30–35 tons of gold reserves but had been closed in 1966. A partially state-owned company, the Société de Recherches et d'Exploitation Minières du Burkina (SOREMIB) was set up to run the mine.[59] Other veins have subsequently been worked in the north (by the Société des Mines de Guiro, or SMG, in the Seno and Namentenga Provinces) and in the Sebba region in a joint venture with North Korea, the Société Minière Coréo-Burkinabè (SOMICOB).[60]

Gold mining has since had a marked impact on Burkina's growth and export receipts. National marketed production of gold (including the marketed portion of artisanal production) reached 3.2 tons in 1986, 3.6 tons in 1987, and 3.7 tons a year later. In 1989, however, the collapse of a mine shaft at Poura kept production at only 2.2 tons. Output resumed its expansion thereafter, with 2.5 tons in 1990, 2.9 tons in 1991, and 3.5 tons in 1992.[61] From 1986 to 1989 the exploitation of gold yielded average annual export revenues of about CFAFr 10 billion or 27 percent of all merchandise exports, which placed gold second to cotton in foreign exchange earnings (cotton accounted for an average 43 percent of the domestic value of merchandise exports over the same period). Most important, gold has brought about a much-needed measure of economic diversification and an opportunity to hedge against climatic adversity. The windfall from mining cur-

*Gold field in the Aribinda region of the Sahel. Courtesy of the Centre National de la Recherche Scientifique et Techniqiue.*

rent known reserves is expected to last until the end of the century. Although new deposits were found in the early 1990s in the provinces of Yatenga, Passoré, and Bam, their potential for commercial or small-scale development has yet to be assessed. In addition to the industrial exploitation of gold, panning by individuals accounts for a large but unknown percentage of national production. It is estimated that the state marketing company, Comptoir Burkinabè des Métaux Précieux (CBMP), which has a formal monopoly on gold purchasing and marketing, controls only about 40 percent of artisanal production, much of the rest being smuggled out of the country.

There is an estimated deposit of 4.5 million tons of zinc with a metal content of 17 percent at Perkoa, near Koudougou. In January 1990 a Swedish company, Boliden International Mining, began exploration work in a joint venture with Burkina's government (which held a 35 percent interest). Production of zinc (expected at 1.3 million tons of 55 percent zinc concentrate over ten years) was scheduled to begin in 1994, but there were reports of the venture's failing financial health at the time.

Reserves of 17.7 million tons of manganese (51 percent content) were first discovered at Tambao, some 322 km north of Ouagadougou, in 1959. Spread over two hills covering a distance of 2.4 km, 83 percent of the reserves have been proven. Studies on the possible exploitation of this resource and the construction of a railroad to reach the area have been undertaken on many occasions since 1960: The Tambao project, truly the phoenix of Burkina's development policy, was resurrected by almost each regime after it was shelved by a discouraged predecessor. According to a 1984 report, there is allegedly enough manganese for production of 350,000 tons a year for thirty-seven years or 540,000 tons a year for twenty years.[62] Despite the abundance of resources, no government ever managed to convince the donor community or foreign companies to invest the necessary amounts in infrastructure to develop the link between Ouagadougou and Tambao that would make production and marketing possible.

The reason for the international community's cold feet was the bleak prospect for the price of manganese on world markets. The Sankara regime decided to bypass these cost-benefit considerations and to rely on "voluntary" labor to extend the rail line, which by 1988 reached Kaya, still less than one-third of the way. Because of the limitations of voluntary labor and the continued restrictions on foreign financing (the World Bank allegedly pressured the government to put its scarce foreign resources to better use), the plan was again shelved in 1990. Nevertheless, existing road links to the Tambao region seemed sufficient to the Canadian company Interstar Mining, which in 1991 entered into an agreement with Burkina for the exploitation of Tambao's manganese. They jointly set up the Compagnie Minière de Tambao (COMITAM), 85 percent of which was owned by Burkina's government. Production and commercial shipments finally began in May 1993, with an early monthly output of about 6,000 tons. The 1993 estimated plant capacity of 140,000 tons per annum was expected to climb to 250,000 tons by late 1994 following scheduled road improvement works.[63] The final arrangement thus provides for road transportation from Tambao to Kaya, from which some of the ore is sent by rail to Abidjan and some by road to Ghana. Sectoral observers expected the foreign exchange earnings impact to be $14 million a year, which would be equivalent to about 5 percent of the 1990 level of merchandise exports.

As a by-product of the development of Tambao, there are plans to exploit the limestone of Tin Hrassan (discovered in 1967), only 35 km away, to supply a proposed cement factory in Ouagadougou. The National Assembly also approved a new, more liberal mining code in 1993 according to which investors will be granted certain incentives, such as exemptions from custom duties on imports of capital goods and fuel and temporary preferential tax treatment.

## *Industrial Underdevelopment*

Of all sectors, industry remains the smallest contributor to Burkina's economy and betrays the country's failure to fully develop away from its agricultural dependency. Furthermore, a substantial part of industrial output is accounted for by extractive activities, which really ought to be considered part of the primary sector, together with agriculture, as most involve the exploitation of underground resources. Out of a total industrial output equal to 20 percent of GDP, manufacture accounted for only a modest 12 percent in 1992 and had actually shrunk (in relative size) since 1970, when it stood at 14 percent of GDP.[64]

With about 120 companies the manufacturing sector is quite modest and matches the limited size of the domestic market. Only ten enterprises account for 80 percent of the sector's value added. About one-third of total manufacturing turnover originates in the food industry, one-third in textiles, 16 percent in the chemical subsector, 12 percent in the metal industry, and the rest in paper and wood-related activities.[65] The main actors of the food industry are SOSUCO (sugar processing), the largest industrial employer, with almost 2,000 workers;[66] Sobbra and Brakina (breweries); and GMB (flour milling). In the textile industry SOFITEX (cotton production and marketing), the largest company in the country, and Faso Fani (cotton clothing) represent almost the entire sector, while leather tanning, a traditionally important activity given the country's comparative advantage in livestock, has now only a marginal economic impact. The main chemical companies are SHSB-CITEC (food oil), SAP (tires and inner tubes), and Sofapil (electrical batteries). The largest metal-based manufacturers are SIFA and SAMFA (bicycles and mopeds) and CBTM (household articles and metal roofing materials). Most companies are located in or around Ouagadougou, with the Bobo region a distant second; the whole sector employs only about 8,000 people.[67]

As is readily visible from the above list, most of Burkina's manufacturing sector, with the notable exception of SOFITEX, is geared toward import substitution. This implies that it has benefited from high levels of protection (such as tariffs and quotas), which have in turn distorted relative prices and wages and hampered efficiency. Most companies are at least in part publicly owned. Eighteen of them are currently being privatized or liquidated under the structural adjustment program of 1991.[68]

Although it accounts for a smaller share of GDP than manufacturing, construction has probably had a significant effect on overall GDP growth since the mid-1980s. The Sankara government's schemes, such as the Sourou dam, Ouagadougou's new central market, and numerous housing projects, contributed to a relative explosion of the buliding sector. There is little in common between the big village that Ouagadougou was in the early 1980s and the fairly urbanized city it has become.

Energy is the Achilles' heel of Burkina's industry. The country is almost totally dependent on thermally generated electricity (from imported oil). Its five stations generate 38.9 Mw, while its hydroelectric generation capacity is 15 Mw.[69]

The picture of Burkina's industry would be grossly incomplete, however, if it did not take into account the ubiquitous informal sector. But because of its very informality, little is known about it (at least in terms of "hard" data). The informal sector can be described as unregistered and mostly unregulated artisanal and small-scale enterprises with little or no accounting basis. It is estimated that in the second part of the 1980s about 40 percent of the value added in the secondary sector (industry) was contributed by the informal sector. The breakdown indicated a 66 percent share of value added in mining, 42 percent in manufacturing, and about 20 percent in construction. As for services, the tertiary sector (not in

*A Dagara woman selling pottery at the market in the city of Dano. Courtesy of the Centre National de la Recherche Scientifique et Technique.*

cluding public administration), two-thirds was in the hands of the informal economy (70 percent commerce, 59 percent transportation, and 76 percent services).[70] Approximate as these numbers are, they nevertheless clearly suggest that the informal economy is not a marginal phenomenon but is at the core of Burkina's secondary and tertiary sectors.

The role of the informal sector in mining clearly refers to private, individual panning. In manufacturing some of the more important activities are the production of *dolo* (the traditional sorghum beer), which is monopolized by women named *dolotières;* tailoring; and carpentry.[71] In the tertiary sector informal agents provide technical services (moped repairs, haircuts), transportation (whether intraurban merchandise transportation on trailers pulled by young men or intercity "bush taxis"), and commerce (many nonregulated market activities).

## Services

### Finance

Burkina shares its central bank, the Banque Centrale des Etats d'Afrique de l'Ouest (BCEAO), with the other members of the Union Monétaire Ouest Africaine (UMOA). UMOA is the West African segment of the Franc Zone, which links the currency of fourteen former French colonies and Equatorial Guinea (the CFA franc) to the French franc. Its fixed parity of CFAFr 50:FFr 1 lasted from

1948 until January 1994, when it was devalued to CFAFr 100:FFr 1. The largest domestic commercial bank is the Banque Internationale du Burkina (BIB), owned by the state at 53 percent and by the West African Banque Internationale de l'Afrique de l'Ouest (BIAO), itself part of Meridien. The second major commercial bank is the Banque Internationale pour le Commerce, l'Industrie, et l'Agriculture du Burkina (BICIA-B), 51 percent state-owned and 45 percent owned by the French Banque Nationale de Paris (BNP). The largest development bank is the Banque Nationale du Développement (BND). Other financial institutions are the Caisse Nationale de Crédit Agricole (CNCA), the Banque Arabe-Libyenne du Burkina (BALIB), and the Union Révolutionnaire des Banques (UREBA). As part of the country's adjustment program, banks have been increasingly opened to private capital, and the government has agreed to limit state participation to 25 percent.

## Transportation

When I first went to Burkina Faso in 1983, I flew with a small charter company called Point Air. Passengers were to meet in Paris. A bus took us to Mulhouse near the Swiss border, where we boarded a plane to Ouagadougou. The return flight was to Marseilles; passengers had to connect with a domestic flight to get back to Paris. Circuitous as they were, these flights were quite cheap and were popular with religious personnel, grassroots development workers, and budget tourists, who would then take the train from Ouagadougou to Abidjan or the bush taxi to the tourist regions of the north and the national parks, the Dogon country in Mali, and other spots. In those days Point Air provided one of the only weekly flights to Ouagadougou from Europe, together with UTA (now Air France) and Air Afrique (of which Burkina is a member). Its profits were reportedly reinvested in local development projects, and it also ran a tourist camp at Gorom-Gorom in the north. Point Air ceased operations in 1987. A government-sponsored charter company, Naganagani, replaced it in 1989. In addition, the official national carrier Air Burkina flies the main domestic and regional connections with a couple of Fokker F-18 aircraft.

Built in colonial times, the 1,183-km railway to Abidjan has played a crucial role in alleviating the handicaps of Burkina's landlocked situation and remains today its main link to the ocean. Called RAN (for Régie Abidjan Niger, a vestige of early colonial ambitions to link Abidjan to Niamey), the rail company was administered jointly by Burkina and Côte d'Ivoire until 1987, when it was split between the two countries because of debt disagreements. The Burkinabè segment was taken over by the newly created Société des Chemins de Fer du Burkina (SCFB), a state company that was put up for sale in 1993. Although the rail link to Côte d'Ivoire is old and deteriorating, 106 km were added to the northeast to Kaya as part of a project to reach the manganese deposit of Tambao. Begun under the Sankara regime, the rail link to Kaya—known as la "Bataille du Rail" (the Battle of the Rail)—was completed in 1988. Plans to extend the rail beyond Kaya were shelved in 1990, and it was decided that the link to Tambao would be established by road. Overall, the country counts only about 1,600 km of paved roads, some of which are in poor condition, and the many dirt roads are vulnerable to inclement weather.

The World Bank's IDA allocated $66 million in March 1992 to complete the $463 million transport sector adjustment program. The program includes the rehabilitation of the railway and roads and the restructuring of transport parastatals, including Air Burkina and Naganagani.

### Tourism

Burkina is not a major tourist destination. The appeal of its climate and scenery is limited. So is its infrastructure. There are three national parks with some wildlife: the Pô National Park, the Arly National Park, and the Grand Parc W. None is a major attraction for tourists from outside the region, as they cannot compete with the parks of East and Central Africa. Burkina's tourists are for the most part low-budget travelers attracted by its cultural wealth. Occasional events, however, do bring in foreigners: the Festival Panafricain du Cinéma de Ouagadougou (FESPACO), the international handicrafts fair, and the biyearly National Culture Week. There are twenty-seven hotels in Ouagadougou and eighteen in Bobo-Dioulasso, with 3,610 beds altogether. Burkina is primarily a stopping point on the way to other destinations; the average length of stay in 1986 was 2.5 days.[72]

## Macroeconomic Performance and Policies

Burkina's overall low level of income contrasts with a relatively well-balanced economy that has been growing at appreciable rates. The real growth of GDP in the 1980s averaged 4.2 percent a year, or 1.6 percent per capita, placing Burkina in the top quarter of African countries.[73] In comparison, sub-Saharan Africa as a whole grew by about 1.9 percent in the 1980s,[74] meaning it actually shrunk by 1.1 percent per capita each year.[75] The countries of Sahelian Africa (Senegal, Gambia, Mauritania, Mali, and Niger[76]), similar to Burkina in their natural resources and climatic conditions, saw their real GDP increase by 2.2 percent per year on average and their per capita GDP decrease by 0.8 percent annually, a performance comparable to other UMOA countries. From 1990 to 1994, Burkina's GDP is estimated to have grown at about 3.8 percent a year (despite the deflationary consequences of the 1994 devaluation), suggesting that the trend of the 1980s is at least being maintained. Despite its wide year-to-year swings (e.g., −1.8 percent real growth in 1984, 9.8 percent in 1985, 10 percent in 1986, and 1.8 percent in 1987), Burkina's economic growth is thus remarkable in both the regional and African contexts (see Table 4.4). Nevertheless, so far it has not allowed Burkina to take off, to modernize the structure of its economy and rise above the other low-income countries.

## Production, Prices, and Income Distribution

Although services are the prime source of value added in the economy, the true engine of long-term growth in Burkina remains agriculture (see Table 4.1). And yet its relative importance in domestic value added has continuously fallen since independence, a trend common to most developing countries and actually believed to be a dimension of development. The fall in agriculture's share of GDP was especially dramatic in the 1960s, when it was due to a deliberate policy of industrialization to create a "modern" sector at the expense of the "traditional" one. In the 1970s the agricultural sector stagnated because of the severe Sahelian drought. The resurgence of more favorable climatic conditions for most of the 1980s triggered a slight comeback until 1990, when the secular trend toward relative agricultural decline continued under the double pressure of the development of the mining and manufacturing sector and the blossoming of services. In addition, the quantitative changes in agriculture have been accompanied by a long-term trend that favors export-oriented growth (cotton is privileged over food crops) compared to an earlier trend in favor of subsistence agriculture.

The industrial push of the early 1960s comes out strongly in Table 4.1. So does its stagnation throughout the 1970s and 1980s. The renewed importance of industry since the mid-1980s is due to mining, construction, and the vitality of the informal sector. The thirty-year period from 1960 to 1990 clearly saw a structural change and a diversification of Burkina's economy, as indicated by the ratio of agricultural to industrial output, which fell from 4.4 to 1.3.

One of the most striking feature of Burkina's national accounts, however, has been the near steady growth of the tertiary sector (services), except in the 1980s. Value added in public administration (in fact the sum total of wages in the civil sector) accounts for about half of services. The slowdown of the 1980s seems to have been due to the extreme austerity measures of the CNR in the civil service and of the cold feet of commercial agents faced with political uncertainty and upheaval. Part of the revival of services in 1990 may be traced to Compaoré's return to urban-biased policies (see the section on agriculture).

The analysis of sectoral performances raises the question of the sectoral distribution of income. No data is available on the size distribution of income (e.g., the share of national income owned by each population quintile). But by dividing each sector's share of GDP by the share of total labor force employed in that sector, one gets an estimate of the relative income of workers in that sector. In 1980 the average income of a worker in industry was almost ten times as high as that of a worker in agriculture; incomes in services were more than six times as high as income in agriculture.[77] It should be borne in mind, however, that there are more active people per household in rural settings, families have other sources of incomes (such as crafts), and they receive transfers from abroad in larger quantities.[78]

Despite Burkina's sustained rate of growth in the 1980s, prices have remained under control, as in the rest of the UMOA (see Table 4.4). The growth of the aver-

TABLE 4.4  Burkina, Sahel, West Africa, and Africa's Economic Performances in the 1980s

| | 1980 | 1981 | 1982 | 1983 | 1984 | 1985 | 1986 | 1987 | 1988 | 1989 | Average[a] 1980–84 | Average[a] 1985–89 | Average[a] 1980–89 |
|---|---|---|---|---|---|---|---|---|---|---|---|---|---|
| **GDP growth (%)** | | | | | | | | | | | | | |
| Burkina Faso | −0.1 | 4.5 | 10.0 | 0.8 | −1.8 | 9.8 | 10.0 | 1.8 | 7.4 | −0.2 | 2.7 | 5.8 | 4.2 |
| Other UMOA | 5.9 | 2.5 | 4.2 | −2.0 | −2.2 | 3.3 | 5.7 | 0.1 | 2.8 | 1.5 | 1.7 | 2.7 | 2.2 |
| Other Sahelian Africa[b] | −1.0 | 3.9 | 6.2 | −0.8 | −4.9 | 2.0 | 7.5 | 2.4 | 4.8 | 2.5 | 0.7 | 3.8 | 2.2 |
| Other West Africa[c] | 4.2 | 3.7 | 2.4 | −1.4 | −0.6 | 3.4 | 3.7 | 2.0 | 3.7 | 2.9 | 1.7 | 3.1 | 2.4 |
| Sub-Saharan Africa | 1.8 | 1.6 | 2.6 | 0.1 | −1.1 | 3.8 | 4.6 | −0.3 | 2.3 | 3.1 | 1.0 | 2.7 | 1.9 |
| **Inflation (CPI, %)[d]** | | | | | | | | | | | | | |
| Burkina Faso | | 7.6 | 12.0 | 8.4 | 4.9 | 6.9 | −2.6 | −2.8 | 4.4 | −0.5 | 8.2 | 1.1 | 4.2 |
| Other UMOA | | 14.4 | 11.8 | 6.1 | 5.2 | 3.0 | 3.6 | −2.6 | 0.9 | −1.1 | 9.4 | 0.8 | 4.6 |
| Other Sahelian Africa[b] | | 11.7 | 13.3 | 6.6 | 14.0 | 10.1 | 19.9 | 5.2 | 2.4 | 4.7 | 11.4 | 8.5 | 9.8 |
| Other West Africa[c] | | 25.9 | 13.7 | 28.0 | 20.0 | 12.7 | 19.7 | 23.8 | 12.7 | 19.5 | 21.9 | 17.7 | 19.6 |
| Sub-Saharan Africa | | 17.1 | 14.6 | 11.4 | 10.2 | 6.4 | 8.5 | 7.1 | 8.9 | 11.0 | 13.3 | 8.4 | 10.6 |
| **GDI/GDP ratio (%)** | | | | | | | | | | | | | |
| Burkina Faso | 18.2 | 17.0 | 21.6 | 20.6 | 17.0 | 25.9 | 22.9 | 20.0 | 22.2 | 22.8 | 18.9 | 22.8 | 20.8 |
| Other UMOA | 24.3 | 20.8 | 21.8 | 15.6 | 11.6 | 15.4 | 15.4 | 14.9 | 16.2 | 14.4 | 18.8 | 15.3 | 17.0 |
| Other Sahelian Africa[b] | 26.3 | 23.2 | 23.0 | 14.7 | 14.9 | 16.4 | 17.5 | 16.6 | 17.1 | 16.9 | 20.4 | 16.9 | 18.7 |
| Other West Africa[c] | 24.3 | 22.5 | 22.7 | 17.0 | 15.6 | 17.0 | 17.4 | 18.3 | 18.8 | 18.4 | 20.4 | 18.0 | 19.2 |
| Sub-Saharan Africa | 20.2 | 20.5 | 17.2 | 13.7 | 10.4 | 11.6 | 14.6 | 15.6 | 15.6 | 15.6 | 16.4 | 14.6 | 15.5 |
| **Resource balance (% of GDP)[e]** | | | | | | | | | | | | | |
| Burkina Faso | −21.3 | −21.1 | −22.5 | −20.8 | −17.7 | −22.7 | −20.0 | −15.8 | −14.9 | −17.6 | −20.7 | −18.2 | −19.4 |
| Other UMOA | −14.1 | −16.1 | −14.7 | −8.6 | −6.2 | −9.8 | −8.2 | −6.8 | −7.7 | −6.8 | −11.9 | −7.8 | −9.9 |
| Other Sahelian Africa[b] | −20.7 | −19.7 | −19.0 | −16.0 | −14.8 | −16.1 | −11.3 | −9.1 | −10.1 | −9.7 | −18.0 | −11.3 | −14.6 |
| Other West Africa[c] | −15.2 | −15.5 | −16.6 | −12.3 | −10.9 | −11.6 | −9.9 | −8.9 | −11.0 | −11.0 | −14.1 | −10.5 | −12.3 |
| Sub-Saharan Africa | 1.0 | −6.2 | −7.0 | −4.4 | −0.2 | 0.1 | −4.0 | −3.3 | −4.0 | −3.0 | −3.4 | −2.8 | −3.1 |

TABLE 4.4 (continued)

| | 1980 | 1981 | 1982 | 1983 | 1984 | 1985 | 1986 | 1987 | 1988 | 1989 | Average[a] 1980–84 | Average[a] 1985–89 | Average[a] 1980–89 |
|---|---|---|---|---|---|---|---|---|---|---|---|---|---|
| Total external debt per capita (in US$) | | | | | | | | | | | | | |
| Burkina Faso | 49.1 | 49.7 | 52.3 | 57.2 | 57.3 | 69.7 | 85.0 | 107.0 | 105.7 | 88.6 | 53.1 | 91.2 | 72.2 |
| Other UMOA | 304.9 | 323.5 | 348.9 | 344.8 | 340.5 | 392.4 | 443.7 | 518.1 | 499.9 | 511.0 | 332.5 | 473.0 | 402.8 |
| Other Sahelian Africa[b] | 265.8 | 304.0 | 332.3 | 354.0 | 367.5 | 407.0 | 469.3 | 542.5 | 524.7 | 514.5 | 324.7 | 491.6 | 408.2 |
| Other West Africa[c] | 256.6 | 289.6 | 316.3 | 330.5 | 334.8 | 380.1 | 430.9 | 501.2 | 626.6 | 489.1 | 305.6 | 485.6 | 395.6 |
| Sub-Saharan Africa | 159.3 | 177.4 | 189.2 | 207.1 | 209.9 | 235.5 | 269.2 | 318.0 | 314.3 | 316.2 | 188.6 | 290.7 | 239.6 |

[a]Averages are simple, unweighted means.
[b]Gambia, Mali, Mauritania, Niger, Senegal.
[c]ECOWAS (Benin, Cape Verde, Cote d'Ivoire, Gambia, Ghana, Guinea, Guinea-Bissau, Liberia, Mali, Mauritania, Niger, Nigeria, Senegal, Sierra Leone, Togo).
[d]Because of incomplete data, inflation sample size varies from year to year.
[e]Resource balance is the difference between exports (fob) and imports (cif) of goods and nonfactor services.

SOURCE: World Bank and UNDP, *African Development Indicators* (Washington, D.C.: World Bank, 1992).

age annual consumer price index in the 1980s was 4.2 percent for Burkina and 4.6 percent for the UMOA as a whole. Other Sahelian countries, however, recorded 9.8 percent average annual inflation and West Africa as a whole 19.6 percent. Because the joint local currency, the CFA franc, is pegged to the French franc, inflation in French and CFA countries converges.[79] Franc Zone countries indeed appear to "import" the French rate of inflation in the long run and to be relatively immune to the inflationary consequences of fiscal policies financed by credit extension or domestic monetary expansion. This could be due to the free circulation of capital within the zone and between France and the zone, which absorb the excess credit.[80]

This low-inflation performance—a blessing in the development context—has come under threat from the devaluation of the CFA franc that took place (mainly at the initiative of France) in January 1994 and changed the peg from CFAFr 50:FFr 1 to CFAFr 100:FFr 1. Although the devaluation immediately led to certain price hikes, it was nevertheless kept at a remarkably low annual rate of 28 percent in 1994. It is as yet unclear, however, whether it will have an inflationary effect in the long run or whether it will encourage the substitution of the consumption of local goods for imported ones and stimulate export revenues.

## Investment, Saving, and Consumption

Burkina has a tradition of active government involvement in development policy. The country has had several five-year and shorter plans that are essentially revolving public investment programs. They identify priority investments in each sector and sources of financing (domestic or foreign, the latter split between loans and grants). Typically, however, most of the financing comes from abroad because of the low level of domestic saving. The 1991–1996 plan, for example, forecasts total investments at around $1.74 billion, which is equivalent to expected total aid flows over the same period.[81] But as elsewhere in Africa, Burkina's plans tend to be excessively optimistic. For example, the 1986–1990 five-year plan achieved only about 60 percent of its targeted aims.[82] Obstacles typically occur when the financing does not come through, where projects are ill defined, or where they suffer from poor implementation and follow-up. The achievement rates give a better indication of the policy priorities than do the preimplementation statements.

Nevertheless, Burkina's active public investment policy has been a significant factor in the country's growth. The share of GDP allocated to investment in the 1980s was somewhat higher than for UMOA countries, especially during the decade's second half, and considerably higher than for Africa in general (see Table 4.4). Investments in mining and in construction infrastructure were likely responsible for the surge observed from 1985 onward. About two-thirds of all investment was accounted for by the public sector. Comparing these public investments with the moderate growth of the share of public consumption, Burkina's government appears thus to provide more of an accumulation impetus than a consumption drain to the economy. As such it offers a counterweight to the emphasis on consumption traditionally present in Mossi and other Burkinabè societies, which is visible in the high level of private consumption and low level of domestic savings. The corollaries to Burkinabè's consumption bias and savings deficiency, however, are the dependence of the government on foreign financing for capital spending and the country's structural trade deficit (the excess of domestic consumption over domestic production).

## Foreign Trade and Payments, Debt, and Aid

Burkina has a substantial structural trade deficit. There has been no instance of a trade surplus at least since before 1970. The deficit nearly doubled between 1973 and 1975 because of the additional food imports required to face the drought. After further substantial increases in 1985 and 1986 (due to the collapse in the price of cotton and rising food imports), the deficit has remained somewhat stable at around $280 million since the second part of the 1980s, thanks mostly to an increase in exports of gold and cotton.

The main products currently exported are cotton, gold, hides and skins, live animals, machinery and transport equipment, vegetables, rubber, groundnuts, sesame seed, oil and fats, and *karité* nuts. The main imports are manufactures, foodstuffs, chemicals, road vehicles and transport equipment, machinery, refined petroleum products, and electrical equipment.[83] Most of Burkina's exports are purchased by France, Taiwan (cotton), Thailand, Portugal, Côte d'Ivoire, and Italy. The bulk of its imports comes from France, Côte d'Ivoire, the United States, Japan, Italy, and Germany. Trade in services is also systematically negative, not least because of the high costs of insurance and freight brought on by Burkina's landlocked situation and because of interest payments on foreign debt.

However, Burkina's overall current account balance displays much lesser deficits than its trade in goods and services and sometimes even records a surplus, such as in 1989 ($71.3 million), or a near equilibrium, as in 1984 (−$3.4 million). The performance of the current account is better than might be expected because of the inflows of private and official transfers, the former being remittances from Burkinabè working abroad and the latter foreign aid grants. Workers' remittances from abroad averaged $37 million a year in the 1970s and $108 million a year in the 1980s, more than the value of agricultural exports. Net official transfers averaged $88 million a year in the 1970s and $224.8 million in the 1980s.[84]

Setting aside the inflows from private and public transfers, which are essentially independent from the performance of the domestic economy, Burkina's external imbalances are generally worse than those of other regional and African countries. The structural deficit of the current account of the balance of payments[85] represents more than twice the average of other UMOA countries. Burkina suffers here from a double predicament. On the one hand, the price of its principal exports (cotton and gold) collapsed during the 1980s. On the other hand, the cereal deficits of the middle 1980s brought about substantial foodstuff imports. The vulnerability of Burkina's economy thus becomes most obvious by looking at its external accounts.

The deficit of the current account is financed mainly by long- and short-term borrowing and by changes in reserves. Foreign direct investments are typically negligible or nil. Nevertheless, despite its need for foreign loans, Burkina has maintained a very moderate level of foreign debts, their burden on the economy remaining quite bearable. Its average debt per capita in the 1980s amounted to $72 as against $403 for the rest of UMOA countries and $240 for Africa in general (see Table 4.4).[86] Its total external debt had reached $956 million or 35 percent of GNP in 1991, whereas many African countries have a level of indebtedness well over 100 percent of GNP. On average, in the second half of the 1980s debt servicing amounted to 8.5 percent of exports of goods and services, representing thus a limited toll on export revenue.[87]

One reason Burkina has managed to keep a low level of indebtedness is that it has avoided white elephant development projects—those heavy in capital requirements and poor in economic returns. Another reason is that its low-income status has allowed it to benefit from a high level of grants and very concessional loans in its development aid. In the 1980s grants made up an average of 77 percent of total disbursed gross official development assistance. The main bilateral donors and creditors are France (with an average annual disbursement of $55.7 million), Germany ($27.8 million), the United States ($26 million), the Netherlands ($24.1 million), and Italy ($14.1 million). Most disbursements from multilateral agencies come from the World Bank ($18.6 million), the European Community ($16.9 million), and the United Nations Development Program (UNDP) ($10.6 million).[88]

## Understanding Burkina's Performance

In view of the economic burden in Burkina's colonial past, its apparent cultural bias against savings, accumulation, division of labor, and technological innovation, its poor climate, few natural resources, and landlocked situation, it seems a bit of a miracle after all that its economy has been performing relatively well—and certainly better and more consistently than those of similar countries. It is the paradox of growth amid poverty.

Some explanations for this are to be found in institutional constraints, which have prevented successive governments from going astray with their fiscal and monetary policies. Burkina's participation in the UMOA, the Franc Zone branch for West Africa, has imposed a fiscal and monetary discipline that has benefited the country's public finances. Yet this variable cannot by itself account for Burkina's record, as other UMOA countries, especially Senegal and Côte d'Ivoire, have experienced dramatic economic imbalances.

Another possible factor lies in Burkina's very poverty. With few resources, there have been fewer opportunities for mistakes and indebtedness. Côte d'Ivoire had a more promising future than Burkina in the 1960s and 1970s because it had greater natural resources. In the 1990s, however, the "Ivorian miracle" has turned sour amid commodity price collapse and foreign debt. In Burkina the impact of the fall in the price of cotton in the 1980s was only as bad as the importance of cotton revenue—quite weak in comparison to cocoa and coffee in Côte d'Ivoire.

Third, just as culture augured ill for the development potential of the peoples of Burkina, culture—albeit another dimension of it—may hold part of the answer of the paradox of growth amid poverty. There is no quantitative or systematic data to support this claim, but anecdotal evidence—starting with the colonial labor policy—abounds in favor of national attributes of hard work, probity, and managerial competence. In its 1989 economic memorandum on Burkina, the World Bank was atypically generous with praise:

> Wherever they may come from, foreign missions visiting Burkina invariably report being favorably impressed by the quality of public sector management, the competence of the officials in charge, the reliability of accounting documents, the regularity with which such documents are produced and, consequently, the speed with which an update of the current situation can be obtained. Other characteristics of Burkina are its sense of realism, its determination to avoid prolonged budget deficits, and the government's capacity to respond rapidly and decisively to threats of financial slippage. . . . These qualities of order and discipline are certainly exceptional assets. . . . Burkina has managed to avoid most of the major investment and management errors which in many African countries have led to large-scale disasters.[89]

Yet the problem with using cultural explanations for economic performance is accounting for the origins of the cultural features in question. The Burkinabè capacity for hard work may have its origins in the harshness of their environment, where subsistence and survival can never be taken for granted. A possible explanation for its managerial competence could be found in the Mossi's experience with statehood and public accountability, which is a trademark of good governance. The checks and balances of the Mossi system together with the practice of large-scale public administration over several centuries may have laid the groundwork for today's relatively efficient and responsive public service and government. In a shorter-run perspective, the personal austerity and ethics of Presidents Lamizana and Sankara have set powerful examples.

In addition, whether "democratic" or dictatorial, Burkina's regimes have always entertained a measure of popular participation that has increased public accountability. Before committing to the structural adjustment program with the World Bank, for example, the government brought together some 2,000 delegates from political parties, trade unions, and other social and professional groups (including agricultural producers and traditional groupings, such as churches and ethnic authorities) in May 1990 to answer the question, "Does Burkina need a structural adjustment program?" Delegates voted in favor of a program that would not reduce the overall wage bill, and the conference gave the government a mandate to continue negotiations with the World Bank and the International Monetary Fund (IMF) "while preserving the major gains of the people."[90] Negotiations led to the signature in March 1991 of a structural adjustment facility (SAF) with the IMF for an amount of $31 million to be disbursed in three installments, conditional upon program implementation, and in June it embarked on a World Bank–sponsored structural adjustment program worth $80 million. The SAF's targets were a GDP rate of growth of 4 percent, an inflation of no more than 4 percent, the elimination of all fiscal arrears, and a reduction in the current account deficit. The World Bank program concentrated on public resource management and private-sector incentives. Although the growth and inflation targets of the SAF were met, the government did not successfully address the issue of fiscal arrears, and the IMF did not disburse more than the first tranche of its SAF. On 1 April 1993, however, it approved a $67 million enhanced structural adjustment facility (ESAF) for 1993–1995 to replace the SAF on more generous terms.

In view of Burkina's performance, the question can be raised whether Burkina's economy actually was in need of such structural adjustment. After all, Burkina's past record was in line with the IMF growth and inflation targets and the program represents a major increase in indebtedness. Indeed, total gross foreign aid disbursements jumped from $354 million in 1990 to $447 million in 1991 and $462 million in 1992.[91] Although grants increased in a similar proportion to total disbursements (maintaining a grant-to-total-disbursement ratio of 75 percent for the 1990–1992 period), the absolute value of loans rose quite significantly.[92] Thus overall indebtedness increased as a result of the adoption of the structural adjustment program.

In fact, the decision to accept a program arose from pressure from three directions. First, the World Bank and the IMF themselves put considerable pressure on most African governments to embark on these flagship programs, and Burkina was no exception.[93] Indeed, the authors of the 1989 economic memorandum, initiated by the bank, make no secret that it was written from the perspective of leading toward adjustment lending. Second, the Compaoré government's more liberal fiscal policies had triggered the problems of arrears accumulation. Finally, under the Sankara government the country had been relatively starved of foreign capital inflows, and there was a domestic demand for money.

It is not necessarily an unreasonable contention that Burkina could have got rid of its arrears and reduced its fiscal and foreign imbalances without resorting to the mixed blessings of adjustment lending. But that is the topic of another book.

## Coping with Poverty: Migration

Migration is central to Burkina's economy. It plays an essential role in the balance of payments, employment, and the generation and distribution of income.

### Colonial Migration

Migration originated under colonial rule as a by-product of high taxation, conscription, and forced labor. Some people emigrated to raise the cash necessary to pay taxes or simply to avoid the heavy burden of French rule and conscription, while others were forced to migrate by the colonial authorities in order to work in large French development projects or plantations.

Most of the "voluntary" migrants went to the Gold Coast. Many of these manual laborers, porters, timber workers, gold miners, and agricultural laborers became permanent migrants; others remained seasonal, coming to harvest cocoa and kola during the dry season in Upper Volta. In 1930 it was estimated that

200,000 Voltaics worked in the Gold Coast.⁹⁴ French anthropologist Guy Le Moal spent some time with migrants in the region of Kumasi in Ghana toward the end of the colonial period. He recalls how the Mossi preferred to work in plantations rather than urban projects because they were more familiar with working the land and could save on housing by living *en brousse* (in the bush). The typical Mossi worker signed a contract with an Ashanti landlord according to which two plots were determined. In one of them the Mossi migrant would grow cocoa; in the other he could grow his own food. Two-thirds of the cocoa output went to the landlord, one-third to the migrant. With good cocoa prices, the migrant could save and after a few years buy the land, turning into a permanent settler. In such cases the workers sent for their families from Upper Volta, and entire neighborhoods were reconstituted in Ghana in which Mossi language and traditions were observed. From October to March, seasonal migrants joined the settlers to work at the peak of cocoa production and returned north in April in time to prepare their plots for the rainy season in Upper Volta.⁹⁵

Forced migrants mostly worked in plantations and on infrastructure projects in Côte d'Ivoire and Soudan. In an effort to redirect migration away from Ghana and into Côte d'Ivoire, France sponsored the creation of Voltaic villages in Côte d'Ivoire on the Ghanaian model in the 1930s. But the scheme met with little success, for until the Front Populaire came to power in France in 1936, the condition of migrant workers in Côte d'Ivoire remained precarious.

## Patterns of Contemporary Migration

Emigration has endured beyond colonial times to become a permanent feature of Burkina's economy and society. The 1985 census assessed the total population of emigrants (defined as having resided or intending to reside abroad for more than six months) at 749,220 (69.2 percent men and 30.8 percent women).⁹⁶ This was more than twice the number of emigrants in the 1975 census (334,755) but nevertheless probably remained an underestimate: Censuses in Côte d'Ivoire and Ghana in 1975 and 1976 suggested that as many as 1 million Burkinabè resided in these two countries. Current total emigrant population is often estimated at between 1 and 2 million (or 10 to 20 percent of resident population), although there are signs that the annual flow of emigration may be drying up. Most emigrants are aged twenty to thirty-five and come primarily from the provinces of Yatenga, Boulgou, Houet, Sanmatenga, Oubritenga, Sourou, and Kadiogo.⁹⁷

Migrants generally go to Côte d'Ivoire and Ghana, language and currency barriers making Ghana the second rather than top choice. Mali is the third most frequent destination, and some migrants travel as far as Gabon. Most migrants work in coastal cocoa plantations and in the industrial sector, though recent trends toward urban employment have been observed.

Today the main reasons for migrating remain economic: to find work, make money, and escape poverty. Among 300 returned migrants interviewed for one study, 40 percent cited the search for money to pay taxes as their reason for migrating, 20 percent said "money" without further qualification, 11 percent wanted clothes and other consumer goods, and 10 percent specified wanting a bicycle (certainly the most popular means of transportation in Burkina).[98] Sidiki Coulibaly, Joel Gregory, and Victor Piché also found employment and money (to pay taxes and buy manufactured goods) to be the paramount reasons Burkinabè gave for migration;[99] and Reardon, Matlon, and Delgado have stressed how migration contributes to a risk mitigation strategy by diversifying household income.[100]

The shortage of land to cultivate due to population pressure in the central Mossi plateau represents one of the roots of poverty and another reason for rural dwellers to leave. The search for a mate appears to be a key motive among migrant women.[101] Marriage is also a reason for male migration—not so much to find a mate as to make the money necessary to pay the expenses of marriage.[102]

Although it drains local labor and human resources, migration has many beneficial effects on Burkina's economy. First, it provides employment to people who would otherwise be unemployed or underemployed (particularly in agriculture, where employment has seasonal dimensions). Second, it relieves the population pressure in the Mossi plateau. And third, it allows for a substantial level of remittances from abroad, which have consistently represented a significant contribution to the balance of payments (with an average of 15 percent of imports of goods and services in the 1970s and 18.4 percent in the 1980s) and thus to national income. In addition to contributing to income growth, remittances from emigrants abroad improve income distribution: Most remittances supplement incomes of rural households, compensating for the country's economic urban bias. Yet the worsening condition of Côte d'Ivoire's economy since the mid-1980s has meant a relative drying up of foreign remittances and a lowering of the beneficial impacts of emigration.

## Notes

1. *The World Development Reports,* published by the World Bank (Oxford: Oxford University Press, various years since 1978), give an annual ranking of the countries of the world in terms of their GNP per capita. Since 1985 Burkina has ranked anywhere from fifth to twentieth poorest country in the world. Variations in the exchange rate between the dollar and the French franc (to which Burkina's currency, the CFA franc, is pegged) are the greatest factors in value and ranking changes. Recent attempts to find a better basis for intercountry comparisons have led to the use of a GNP per capita expressed in "purchasing power parity." This takes into account and compensates for the much lower valuation of some nontradables in developing countries that tend to depress traditional GNP figures. According to this new method, Burkina's GNP per capita jumps from $300 to $730 in 1992; *World Development Report 1994* (Oxford: Oxford University Press, 1994), 162, 220.

2. Catherine Coquery-Vidrovitch, ed., *L'Afrique occidentale au temps des Français. Colonisateurs et colonisés, 1860–1960* (Paris: La Découverte, 1993), 263.

3. Louis Tauxier, *Le Noir du Soudan* (Paris: Larose, 1912), 538, quoted in Elliott P. Skinner, *The Mossi of Burkina Faso: Chiefs, Politicians and Soldiers* (Prospect Heights, Ill.: Waveland Press, 1989), 159.

*The Economy of Growth amid Poverty*

4. Coquery-Vidrovitch, *L'Afrique occidentale,* 266–267.

5. Elliott P. Skinner, *African Urban Life: The Transformation of Ouagadougou* (Princeton, N.J.: Princeton University Press, 1974), 29.

6. Ibid., 32.

7. Coquery-Vidrovitch, *L'Afrique occidentale,* 285.

8. Alan P. Fiske, *Structures of Social Life. The Four Elementary Forms of Human Relations: Communal Sharing, Authority Ranking, Equality Matching, Market Pricing* (New York: Free Press, 1991).

9. Ibid., 233.

10. Ibid., 265.

11. Ibid., 269.

12. Ibid., 272–273.

13. Mahir Saul, "Money and Land Tenure as Factors in Farm Size Differentiation in Burkina Faso," in R. E. Downs and S. P. Reyna, (eds., *Land and Society in Contemporary Africa* (Hanover, N.H.: University Press of New England, 1988).

14. Fiske, *Structures,* 268.

15. Ibid., 247.

16. See Richard Vengroff, *Upper Volta: Environmental Uncertainty and Livestock Production* (Lubbock: International Center for Arid and Semi-Arid Land Studies, Texas Tech University, 1980), 25–26.

17. See Hans Binswanger, John McIntire, and Chris Udry, "Production Relations in Semi-arid African Agriculture," in Pranad Bardhan, ed., *The Economic Theory of Agrarian Institutions* (Oxford: Clarendon Press, 1989), 122–144; and Marcel Fafchamps, "Solidarity Networks in Preindustrial Societies: Rational Peasants with a Moral Economy," *Economic Development and Cultural Change,* 41, 1 (October 1992): 147–174.

18. See also Mamadou Dia, "Development and Cultural Values in Sub-Saharan Africa," *Finance and Development,* December 1991, 10–13.

19. Fiske, *Structures,* 268.

20. Titinga Frédéric Pacere, *Ainsi on a assassiné tous les Mossé: Essai-témoignage* (Sherbrooke, Quebec: Editions Naaman, 1979), 113.

21. Fiske, *Structures,* 340–341.

22. Claudette Savonnet-Guyot, *Etat et sociétés au Burkina: Essai sur le politique africain* (Paris: Karthala, 1986), 35–39.

23. Jacques Lecaillon and Christian Morrisson, *Economic Policies and Agricultural Performance: The Case of Burkina Faso* (Paris: Development Centre of the OECD, 1985), 21.

24. World Bank, *Burkina Faso: Economic Memorandum,* vol. 1 (Washington, D.C.: World Bank, 1989), 52–53.

25. Unfortunately, good harvests are also occasionally threatened by locusts, which can devour a substantial percentage of the crops in a short time.

26. I have calculated the correlation coefficient based on rainfall data from the Institut Météorologique National, cereal output data from the Institut National de la Statistique, and GDP data from the IMF, the Banque Centrale des Etats d'Afrique de l'Ouest, and the Institut National de la Statistique. The data covered the years 1980 to 1989 inclusive. I also found a correlation coefficient of $r = .49$ between Lecaillon and Morrisson's rainfall index and IMF real GDP figures for 1972 to 1981.

27. Economist Intelligence Unit (EIU), *Country Profile: Niger, Burkina Faso* (London: EIU, 1993–1994), 37.

28. Ibid., 39.

29. Lecaillon and Morrisson, *Economic Policies*, 21.

30. Ibid., 26.

31. Ibid., 21.

32. Ibid., 26.

33. Jacqueline Sherman, *Grain Markets and the Marketing Behavior of Farmers: A Case Study of Manga, Upper Volta* (Ann Arbor, Mich.: Center for Research on Economic Development, 1984), quoted in Fiske, *Structures*, 265.

34. Lecaillon and Morrisson, *Economic Policies*, 33, 46.

35. World Bank, *Social Indicators of Development 1994* (Oxford: Oxford University Press, 1994), 53.

36. Ibid., and World Bank, *Economic Memorandum*, 64.

37. EIU, *Country Profile*, 47.

38. Thomas Reardon, Peter Matlon, and Christopher Delgado, "Coping with Household-Level Food Insecurity in Drought-Affected Areas of Burkina Faso," *World Development*, 16, 9 (1988): 1065.

39. Ibid., 1069.

40. See also Thomas Reardon, Peter Matlon, and Christopher Delgado, "Determinants and Effects of Income Diversification Amongst Farm Households in Burkina Faso," *Journal of Development Studies*, 28, 2 (January 1991): 264–296.

41. Production contracted somewhat to 167,170 tons in 1991–1992, 163,000 in 1992–1993, and a poor 116,000 in 1993–1994 because of a regional cyclical downturn throughout West Africa.

42. Index figures until 1980 were derived by Lecaillon and Morrisson, *Economic Policies*, 48. I computed updates from 1980 onward using the same methodology and relying on data from the IMF's *International Financial Statistics* (various issues) for cotton prices, the dollar-CFAFr exchange rate, and Burkina's CPI.

43. See Reardon et al., "Coping," 1070.

44. The majority of Burkina's Peuls live in permanent settlements, migrating to and from their homes as they accompany cattle.

45. See Vengroff, *Upper Volta*.

46. Ibid., 59.

47. Ibid., 60.

48. Ibid., 61.

49. Lecaillon and Morrisson, *Economic Policies*, 17.

50. EIU, *Country Profile*, 39.

51. Lecaillon and Morrisson, *Economic Policies*, 18.

52. EIU, *Country Profile*, 39.

53. Lecaillon and Morrisson, *Economic Policies*, 19.

54. Bernard Tallet, "Le CNR face au monde rural: Le discours à l'épreuve des faits," *Politique Africaine*, 33 (1989): 39–49.

55. World Bank, *Economic Memorandum*, 60.

56. World Bank, *Annual Report 1991* (Washington, D.C.: World Bank, 1991), 142.

57. World Bank, *Economic Memorandum*, 65.

# The Economy of Growth amid Poverty

58. See Jean-Baptiste Kiéthéga, *L'Or de la Volta noire: Archéologie et histoire de l'exploitation traditionnelle* (Paris: Karthala, 1983).

59. The Burkinabè state owns 60 percent of SOREMIB. The other shareholders are the Islamic Development Bank, the European Investment Bank, and French investors.

60. The SMG was liquidated in July 1993 for lack of profitability.

61. Secrétariat de la Zone Franc, *Rapport Annuel* (Paris: Secrétariat de la Zone Franc, 1993), and EIU, *Country Report*, 4, 1994.

62. Hama Dicko, "Essai d'étude socio-économique du projet minier-ferrovière de Tambao. Les enjeux, les impacts, les répercussions du projet de développement des transports sur la base d'une voie ferrée Ouagadougou-Kaya, Dori, Tambao, Tin-Hrassan sur le développement socio-économique de la région et de la nation." Université de Ouagadougou, 1984. Mimeographed.

63. EIU, *Country Profile*, 52.

64. World Bank, *World Development Report 1994*, 166.

65. World Bank, *Economic Memorandum*, 80.

66. See Taladidia Thiombiano, *Une enclave industrielle: La société sucrière de Haute-Volta* (Dakar: Codesria, 1984).

67. EIU, *Country Profile*, 41.

68. See ibid., 42, for a full list of the companies scheduled for privatization and/or liquidation.

69. Ibid., 41.

70. World Bank, *Economic Memorandum*, vol. 1, 3.

71. According to Meine Pieter van Dijk, *dolo* making is the largest branch of the informal sector in Ouagadougou in terms of employment. See Meine Pieter van Dijk, *Burkina Faso: Le secteur informel de Ouagadougou* (Paris: L'Harmattan, 1986), 105.

72. EIU, *Country Profile*, 45.

73. Based on UNDP and World Bank, *African Development Indicators* (Washington, D.C.: World Bank, 1992), and computed in dollars at constant 1987 prices. I estimated a real rate of growth in domestic currency of 3.7 percent at constant 1985 prices for the same period based on statistics from the UMOA, the IMF, the World Bank country team, and the Economist Intelligence Unit. Both figures fall in the same range. The *African Development Indicators* figure is used here to favor comparisons with other countries.

74. Ibid., 31.

75. Assuming a 3 percent annual population growth rate.

76. Chad was excluded, as its performance was excessively affected by civil war in the 1980s.

77. For data on the economically active population, see International Labor Office, *Economically Active Population Estimates and Projections, 1950–2025* (Geneva: International Labor Office, 1986). For the 1980 ratios, I divided the percentage of GDP in the sector by the percentage employed in the sector. The results were as follows: agriculture: $45/86.6 = 0.52$; industry: $22/4.3 = 5.12$; and services: $32/9.1 = 3.52$.

78. Lecaillon and Morrisson, *Economic Policies*, 27, 28.

79. See Patrick Honohan, *Price and Monetary Convergence in Currency Unions: The Franc and Rand Zones*, Policy, Research, and External Affairs Working Paper Series 390 (Washington, D.C.: World Bank, 1990).

80. There are also substantial disadvantages associated with the Franc Zone for a country like Burkina. For the merits and inconveniences of the Franc Zone (both purely economic and from a political economic point of view), see Shantayanan Devarajan and Jaime

de Melo, "Evaluating Participation in African Monetary Unions: A Statistical Analysis of the CFA Zones," *World Development*, 15 (1987): 483–496; Nicolas van de Walle, "The Decline of the Franc Zone: Monetary Politics in Francophone Africa," *African Affairs*, 90 (1991): 383–405; and Lisa M. Grobar, "Money and Macroeconomic Adjustment in the CFA Zone" (California State University at Long Beach, 1994, mimeographed).

81. EIU, *Country Profile*, 34–35.
82. Ibid.
83. For quantified information, see BCEAO, *Statistiques Economiques et Monétaires*.
84. All balance-of-payments data from IMF, *International Financial Statistics Yearbook* (Washington, D.C.: IMF, 1992).
85. Not including private and public unilateral transfers or factor services exports and imports.
86. The decline of Burkina's total debt and its current account surplus in 1989 are due to French measures of debt forgiveness. Other countries have benefited from similar measures in different years.
87. All debt figures from World Bank, *World Debt Tables* (Washington, D.C.: World Bank, 1992).
88. All aid data are from OECD Development Assistance Committee, *Geographical Distribution of Financial Flows to Developing Countries* (Paris: OECD, 1994).
89. World Bank, *Economic Memorandum*, vols. 1, 16.
90. EIU, *Country Report: Togo, Niger, Benin, Burkina*, 3, 1990.
91. OECD, *Geographical Distribution*, 1994.
92. Ibid.
93. See Paul Mosley, Jane Harrigan, and John Toyle, eds., *Aid and Power: The World Bank and Policy Based Lending* (London: Routledge, 1991).
94. Coquery-Vidrovitch, *L'Afrique occidentale*, 273.
95. Guy Le Moal, *Au Ghana avec les travailleurs voltaïques* (Ouagadougou: Institut Français d'Afrique Noire, 1965). Note that Le Moal refers to savings and accumulation by Mossi migrants. This contrasts with the "traditional" Mossi economic outlook, which stresses consumption and distribution. One possible explanation is that these Mossi underwent a cultural change through their contact with both the "modern" economy in Ghana and the entrepreneurial attitudes of the Ashanti. See also Enid Schildkrout, *People of the Zongo: The Transformation of Ethnic Identities in Ghana* (Cambridge: Cambridge University Press, 1978).
96. Institut National de la Statistique et de la Démographie. *Deuxième recensement général de la population du 10 au 20 décembre 1985: Principales données définitives* (Ouagadougou: Ministère du Plan et de la Coopération, n.d.), 20.
97. Although a phenomenon of smaller proportions, there is also a fair amount of domestic migration in Burkina, mostly from central and northern regions to the west and southwest for agricultural work.
98. Cited in Fiske, *Structures*, 331.
99. Sidiki Coulibaly, Joel Gregory, and Victor Piché, *Les Migrations voltaïques*, vol. 1: *Importance et ambivalence de la migration voltaïque* (Ottawa: Centre Voltaïque de la Recherche Scientifique, Institut National de la Statistique et de la Démographie, 1980).
100. Reardon et al., "Coping," 1065.
101. Coulibaly et al., *Les Migrations*, 77–78.
102. Cited in Fiske, *Structures*, 331.

# 5

# SOCIETY AND CULTURE

Burkina's society is composed of many societies, its culture the fruit of many cultures. The Burkinabè nationality overlaps ethnic identity, French competes with a multiplicity of languages used in everyday life, and the secularism of the state rivals the spiritual powers of animism, Islam, and Christianity. Yet there is also a Burkinabè society and a modern culture sponsored by the state, which finds its roots in—and attempts to blend—the many local ones. Remarkably and somewhat surprisingly, Burkina's cultures and societies provide a human environment in which religious and ethnic conflicts are limited, if they exist at all. This chapter gives a taste of this social diversity and cultural wealth and seeks explanations for Burkina's civil peace, which contrasts both with the upheaval of its polity and with the societies of many other African states.

## Population and Demography

Burkina's population, estimated at 9.5 million in mid-1992, has been growing at about 2.6 percent per year since 1980. This was an acceleration over the previous two decades, when the rate of growth was estimated to have been, respectively, 1.6 percent and 2.1 percent.[1] Starting with a population of about 5 million in 1960, Burkina thus probably reached the 10-million threshold in 1994. This relatively rapid (but average for Africa) population increase is due to a faster decline in mortality than in birthrates, the consequence of improved health standards but relatively unchanged attitudes toward children. Mortality, as assessed by crude death rates, has fallen from twenty-seven per 1,000 in 1960 to eighteen per thousand in 1992. Crude birthrates, in contrast, have only fallen from forty-nine per 1,000 in 1960 to forty-eight per 1,000 in 1992.[2]

Current population figures are estimates based on the second national census, which dates back to 1985 (the first census having taken place in 1975 and another planned for 1995). At that time total resident population amounted to 7.96 million.[3] The most populated province was Yatenga, with more than half a million

*A Wunie (Ko) dwelling in the Boromo region, with traditional door and drying corn to be used for next year's seeds. Courtesy of the Centre National de la Recherche Scientifique et Technique.*

people, followed by Kadiogo (in which the capital, Ouagadougou, is located), with slightly less than half a million. Yet since 1975 the population of Kadiogo had increased by an annual rate of 9.6 percent, while Yatenga had risen by only 0.1 percent a year. This indicates the existence of rural-urban migration in addition to international migration. In 1985 the average village counted 1,117 inhabitants, and only thirty-one towns had more than 10,000 dwellers. Total urban population was estimated at 1,011,074 (or 12.7 percent of total), up from 362,610 10 years earlier, for an urban growth rate of 10.8 percent per year (compared to 3.7 percent per year between 1960 and 1975). The largest towns are the capital, Ouagadougou (with about half a million inhabitants), Bobo-Dioulasso (250,000), Koudougou (55,000), Ouahigouya (45,000), Banfora (40,000), and Kaya (30,000).[4] The overall population density in 1985 was twenty-nine inhabitants per square kilometer. It may have reached 36.5 inhabitants per square kilometer in 1994. The spatial distribution is very uneven, however. In 1985 the provinces of Kadiogo and Kouritenga (where Koupela is the largest town) had a population density above 100 per square kilometer. The rest of the Mossi country stood between thirty-six and 100, while the east (Gourma and Tapoa), the north (Soum and Oudalan), and parts of the southwest (Comoe) had fewer than fifteen inhabitants per square kilometer.[5] The 1995 census should show an increase in the density of the west and southwest relative to that of the Mossi plateau.

In 1985, 51.9 percent of all Burkinabè were women. In the cohort below the age of eighteen, there is a female deficit, which may be due to higher female child mortality. In the twenty- to sixty-four-year-old cohort, there is a male deficit; the deficit is largest among the group aged twenty to fifty-five because of emigration. For the same reason, men are more numerous than women in urban areas.[6]

A total of 48.3 percent of all Burkinabè were less than 15 years old in 1985. Between 1975 and 1985 there has been a simultaneous increase in the proportion of children and elderly. Since both are considered economically inactive (this is only a statistical definition, for anyone who has been in Africa knows that children become effectively active at an early age), the polarization of the population triggered an economic dependence ratio (defined as the ratio of inactives to actives) of 110 percent in 1985.[7]

## Ethnicity and Language

Mossi make up 50.2 percent of Burkina's population.[8] No other ethnic group comes close: The Peul are about 10 percent; Mandingue 7 percent; Lobi, Dagiri, and similar groups 7 percent; Bobo 6.7 percent; Senufo 5 percent; Gurunsi 5 percent; Bissa 4.7 percent; and Gurmanche 4.5 percent (see Map 5.1).[9] In fact, Burkina's sixty or so ethnic groups can be aggregated into three families: The Voltaic, or Gur, family includes the Mossi, Bwa, Gurmanche, Lobi, Senufo, and Gurunsi among others; the Mande family comprises the Samo, Marka, Bobo, Bissa-Boussanse, and Dioula; and the West Atlantic family has the Peuls as a single member. The distinction between Voltaic, Mande, and West Atlantic is based on language.[10] There is thus more linguistic than ethnic homogeneity. One can get by in most areas of Burkina by speaking either Moré, Peul (in the north), or Dioula/Bambara (in the west). In addition, French is the official language throughout the territory. Fewer people speak French in rural areas than in the cities, however. Some English is spoken in the region that borders Ghana.

### *Is Burkina a Mossi State?*

Given the demographic weight of the Mossi and their place center stage of Burkina's precolonial, colonial, and independent history, it is fair to ask whether Burkina is a Mossi state. In other words, do the Mossi exert a hegemonic domination over the other groups through the levers of modern state power? Did they hijack the state that had itself earlier hijacked their own history? Although Burkinabè and foreign observers occasionally maintain that this happened, the answer is ambiguous and complex and deserves special attention.

The factor most often cited in support of the theory of "Mossification" of Burkina is that no other group is so large, so politically organized, and capable of exerting the same influence over the institutions of modern statehood. Savonnet-Guyot makes this point by comparing the Mossi with the "mosaic" of small ethnic groups who live in the south and west and who, because of their lower demographic density, their lack of centralized political and administrative systems, and their unfamiliarity with the institution of statehood are "Ill prepared for the con-

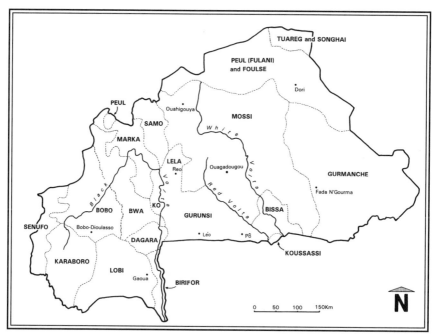

MAP 5.1    Burkina Faso: Ethnic Map

quest of the modern state's political and administrative apparatus."[11] She acknowledges, however, that individual strategies may be more important than collective ones in the acquisition of power, explaining thereby the long joint rule of Lamizana and Saye Zerbo, two Bissa leaders. Nevertheless, beyond the almost accidental ascendancy of non-Mossi elites, she believes in the capacity of the Mossi to progressively colonize the modern state—or at least in their ongoing colonization of territories outside their traditional area:

> [The Mossi] system . . . covers a society whose warring and colonizing past has not encouraged the development of modern democracy but [that has been] powerfully helped by its long expansionistic and assimilationist tradition to master a space that has become "national" [and] of which it is capable of reducing the differences and unevenness. . . . A truly interior colonization of territories that never belonged to the *naam* is currently taking place in the country's west and southwest because of the overpopulation of the Mossi's own domain.[12]

The colonization of the state and the territorial expansion of the Mossi are two distinct propositions, however, and the latter does not necessarily imply or lead to the former. Nevertheless, it should be borne in mind that the Mossi kingdoms were created out of the Dagomba expansion from Ghana because of population pressures and that the Mossi are quite integrative by nature, typically letting refugees and seminomadic peoples settle in their territory and frequently marrying outside their group. That the Mossi plateau is currently undergoing substantial demographic strains is not disputed. There is thus the possibility that history may repeat itself.

Another recurrent argument highlights that the Mossi monarchy was never destroyed by the French. Its institutions survived, unlike those of more Muslim states.[13] Granted, the argument goes, they were in some measure neutralized by the French and perverted by Mossi collaborationism. Nevertheless, when Upper Volta became independent, the *mogho* was still in part alive as a political system, and its king resided only a few kilometers from the presidential palace.

This theory, which equates the Mossi people with the Mossi monarchical structure and its leadership, is difficult to reconcile with one event and one trend in Burkina's history.[14] The event is the failed monarchical coup of the *mogho naba* in 1958. As outlined in Chapter 2, the newly crowned Mogho Naba Kougri tried to impose a constitutional monarchy onto the Territorial Assembly on 17 October 1958 by sending 3,000 armed warriors to surround the parliament. They quickly retreated, however, when faced with the colonial troops. Rather than demonstrating the vitality of the Mossi monarchy, this episode in fact sealed its fate once and for all and made plain the rupture that Voltaic state politics represented with the Mossi system. There was no room for Mossi politics per se in Upper Volta. The new political elites were claiming their emancipation from the *naam*. Without might, the *mogho naba*'s natural right to rule was no more.

This 1958 event was but a manifestation of a larger trend, the intra-Mossi political split that occurred during the 1950s. "Modernized" and formally educated Mossi elites who inherited the colonial state often were not *nakombse* and made it a point to differentiate their legitimacy from that of the ethnic rulers. The reason commoners rather than royals reached political power is that for a long time ethnic chiefs refused to send their children to colonial and missionary schools, which they rightly perceived as threatening their authority and their system. "As a result," Skinner writes, "few of the chiefs' sons received an education, and thus they were later unable to compete effectively with commoners who had gained a valuable education at the missions."[15] The consequences of this failure of ethnic elites to engage in the new avenues of social and political power became apparent as early as 1946. According to Skinner, the 1946 electoral law was "heavily weighted in favor of the educated and urban elite. . . . Candidates for election . . . had to . . . have a good knowledge of French. . . . Many chiefs were eligible to vote, but, because their fathers had opposed their being educated at mission-sponsored schools, few were qualified to run for office."[16] This phenomenon of marginalization of *nakombse* families with respect to education and thereby to administration was thus of their own making. One of the first of these formally educated new elites had been Philippe Zinda Kaboré, who died in Abidjan in May 1947, shortly "after making a violent speech in Ouagadougou denouncing the Moro Naba."[17] Other members of the "new generation" of *naam*-less Mossi politicians were Joseph Ouédraogo and Maurice Yaméogo. Joseph Conombo, in contrast, did

*A Wunie (Ko) mask in the Boromo region. Courtesy of the Centre National de la Recherche Scientifique et Technique.*

attempt to represent the interests of the Mossi hierarchy in the late 1950s. He failed to rise to political prominence, however, until he became prime minister under Bissa president Lamizana in 1978 and was thus unable to influence the political system in the interests of Mossi ethnicity. As a result, even in positions of state power the Mossi have not been agents of ethnic power since the 1950s.

There is no doubt, however, that the Mossi hierarchy wanted to inherit the levers of independent Upper Volta, become the natural rulers of the new state, and extend the effects of the *naam* beyond the Mossi. The 1958 coup attempt attests to this. And its failure did little to change the *mogho naba*'s aspirations. In a January 1959 speech before the Territorial Assembly, the Mossi leader made it clear that "our intention is not to relinquish the country's direction only to elected officials but to have a close collaboration with them because if the representative elected yesterday has rights, the representative of many centuries has also preserved his."[18] As late as 1969, in fact, Mossi chiefs still begged to be integrated into the political system, requesting a constitutional provision for making the *chefferie* a national institution.[19] But this, too, failed. In the end, the desires of the *mogho naba* were frustrated and thus should not be construed as evidence of Mossi ethnic power in the state.

By the end of the colonial period, Mossi ethnicity had all but ceased to be a variable of Voltaic politics. The last time Mossi ethnicity was a relevant political force was in the immediate aftermath of World War II, when, in Skinner's words, "with the reconstitution of Upper Volta, the Mogho Naba and his chiefs attained their highest prestige since the conquest."[20] The Union des Chefs Traditionels, created in 1952 under the chairmanship of the *mogho naba,* had among its objectives "to collaborate closely with the economic and social aims of the French administration."[21] It never really took off.

Consequently, the mere identification of a politician as Mossi is mostly devoid of implication for ethnic politics. That fully or partly Mossi presidents (Maurice Yaméogo, Jean-Baptiste Ouédraogo, Thomas Sankara,[22] and Blaise Compaoré) have ruled Burkina for eighteen of its thirty-four years as an independent country does not imply the "Mossification" of the state. Quite the contrary: Mossi presidents—Yaméogo and Sankara above all—have typically been harshest toward the *chefferie.* As for Compaoré, he seems to want to be associated with it only at election time.

## Relations Between State and Ethnic Powers

Having assessed the relative independence of the state vis-à-vis Mossi ethnicity, we might take the reverse perspective and look at the state's attitude toward ethnic groups, the latter being an essential feature of the civil society to which we referred in Chapter 3.

The contemporary state's view of ethnicity is an extension of the colonial state's attitude. The latter was not without ambiguity but in general sterilized the powers of ethnicity. As an invader aiming to impose its own system, France crushed the political authority of ethnic chiefs, reduced their autonomy, and transformed them into its auxiliaries.[23]

Political authorities in independent Upper Volta and later Burkina have continued the same mix of successive use and abuse of ethnic powers but have seemed to favor the latter even more than did the French colonials. Just as in colonial times, they have used the chiefs to compensate for the shortcomings of the administration, thereby reinforcing their authority. But they have also wanted to limit the chiefs' authority to avoid seeing theirs being threatened. Eventually and ideally, the independent state wishes to bypass the chiefs.

The early independent state of President Yaméogo was still insecure, run as it was by "modernized" Mossi elites who were both afraid of their potential submission to chiefs and eager to show their independence from them. The first republic therefore provided the stage for an all-out assault by the state on the political structures of ethnicity. In 1962 Yaméogo passed a decree forbidding "all signs and all external manifestations of ancient customary hierarchies, . . . all practices of subordination incompatible with the principles of equality and dignity of all citizens."[24] According to another 1962 decree, chiefdoms above the village level left vacant following a chief's death or some administrative measure were no longer to be maintained. Canton and village chiefs later lost their right to a salary. In 1964 a new decree provided for the election of village chiefs by universal secret ballot, further reducing the role of the custom.[25] Otayek summarized Yaméogo's motives:

> in order to impose his authority in the rural areas . . . which recognised themselves (and still recognise themselves) in customary institutions, he had to dismantle them and undermine the foundations of this power . . . [which] testified to the willingness of the state to do without the "relay" represented by the chieftaincy in order to impose its own instruments of domination.[26]

In view of these policies, the traditional chiefs' relief at seeing Yaméogo go in 1966 is not surprising and provides additional evidence, if need be, of the lack of ethnic control over the state. Yet although he somewhat revalorized their status, Lamizana also confirmed the system of election of village chiefs. Nevertheless, his government was tolerant of situations where customary authorities appointed a chief other than the one who had been elected and looked benevolently on the candidacies of brothers, sons, and grandsons of chiefs. As a result, the reform of the *chefferie* was all but actually bypassed. In addition, the administration continued to resort to chiefs as auxiliaries and effectively integrated them into public administration.

The CMRPN and the CSP were not in power long enough to have defined and consistent ethnic policies. The CNR, in contrast, came to power with well-entrenched, ideologically inspired attitudes toward the *chefferie* and wasted little time implementing them. Two decades after Yaméogo, Sankara acted out of much the same rationale, bound as he was by the enduring unsteadiness of the state twenty-three years after independence.

Two months after seizing power, the CNR declared the Committees for the Defense of the Revolution to be the "authentic organization of the people,"[27] denying a reality in which the people organized in households, villages, and ethnic groups. The imposition of CDRs in villages and neighborhoods followed the same objective of replacing the legitimacy of history and custom with a new legitimacy of revolution.[28]

On 4 August 1984 the CNR nationalized the land as part of a sweeping land reform whose objective was to break the power of traditional chiefs. Burkina was in little need for land reform in terms of improved equality and access to land, for it does not have huge estates monopolized by a few landlords nor dispossessed rural laborers in need of their own parcels. But to the extent that control over and management of the land is an essential part of ethnic organization, the nationalization of land by the CNR was a sharp blow to the heart of ethnicity. It perturbed the bonds between *nakombse* and *tengbiise,* defined as they are in terms of their relation to land. Nor did the CNR come up with an alternative to the ethnic organization and management of land and agriculture. There were certainly divisions among its members as to whether a state farm system was orthodox and feasible or whether village cooperatives should take over. As a result, the land reform was announced in an atmosphere of improvisation. In any case, the CNR did not have time to deliver a policy on this matter before being overthrown in 1987. Land reform remained a dead letter, and the ethnic authorities gradually regained their earlier prerogatives.

Nevertheless, both the CDR and the land reform policies were indicative of the CNR's desire to use the state to wage its revolution against chiefs and customs, against age, against history. By painting ethnic authorities as enemies of his revolution, Sankara encouraged the emergence of contradictions, as his ideology demanded. According to Skinner, Sankara was "determined . . . to finally deprive the Mossi chiefs, who were survivors of a decadent feudal past, of the power they had held over the centuries."[29] But his identification of Mossi and other chiefs as feudal was mistaken.[30] The dogmatic thought of the CNR leadership, apparently the result of a poorly digested reading of Marx and Engels's *Communist Manifesto*, was eager to identify Burkina's stage of development according to "scientific socialism." But conditions in Burkina in the 1980s were different from those of England and continental Europe in the Middle Ages that inspired Marx and Engels to define the "feudal relations of property" from which the bourgeoisie would emerge.[31] Burkina's peasants are not vassalized by village chiefs, and the latter can barely restrict access to their land. Intent upon applying a rigid and dogmatic version of Marxism, Sankara failed to see the powerful contradictions of his time, mainly that between the postcolonial state and a peasant civil society. This contradiction may not revolve around ownership of the means of production. After all, the Mossi and other economic cultures, in a situation of low population density and land abundance, have kept Burkina away from the inequality-inducing phase of "primitive accumulation," leaving the country still rather overwhelmingly in the situation Marx described as that of peasant owner-worker.

This antiethnic attitude was not limited to the Mossi but also extended to the eastern and central *chefferies*.[32] The CNR's ethnic policy fits well Bayart's description of state-society relations in Africa: "[Rulers'] objective is to enlist the dominated social groups within the existing space of domination and to teach them to be subject of the state. The aim is to administer society, even against itself, and to order it according to the explicit, ideal canons of modernity."[33] In the case of the CNR, these canons were defined in terms of hypothetical classes.

## *The Absence of Interethnic Conflict*

Most of the conflicts and contradictions that involve ethnicity have been mentioned in terms of its relationship to the state. Thus, although I have made many references to the Mossi, they actually apply to ethnicity in general and not to one group versus another. There is indeed relatively little conflict among ethnic groups in Burkina, unlike in many other African states, where ethnic confrontation has become a social plague. Several explanations can be put forward to shed light on Burkina's relative uniqueness in this respect. First, the assimilationist nature of the Mossi encourages identification over differentiation and blends ethnic distinctions. Similarly, because the whole *mogho* is contained within the territory of Burkina, irredentism has never emerged, and its sheer size within Burkina—both absolute and relative to other groups—has mitigated against the development of strong domestic antagonisms. Second, the deliberate action of the state to distance itself from ethnic authority, if not to crush it, has prevented any one group from seizing the levers of power and has made it unlikely that any would try. By the same token, the root of many ethnic rivalries, the object of ethnic competition—state "ownership"—has been eradicated. Third, the paucity of economic resources available in Burkina and the limited avenues to wealth have made it quite unworthy the effort to compete for them. Finally, many among Burkina's ethnic groups have a nonnationalistic outlook, especially in the west, where identity and the sense of belonging are defined primarily at the household or village levels. This has two consequences: (1) Individuals who do not define themselves by affirming their own ethnicity also do not do so in opposition to other ethnic groups; (2) their lifestyles are not conducive to collective action, which is a prerequisite for ethnic conflict.

All this is not to say that there is no occasional rivalry, jealousy, or accusation of ethnic favoritism. But it does imply that they have never reached a level of concern to the state. This has no doubt been one of the reasons the unsteadiness of the Burkinabè state has not led to its collapse. The absence of strong political ethnic identities has prevented civil society from offering a credible alternative to the Burkinabè state, and the absence of interethnic conflict has saved Burkina from a real test of its statehood. For if Burkina's unstable polity had had to evolve within an environment of ethnic unrest, it could have fallen victim to outright civil war.

Whether this blessing will last has become increasingly questionable over the last decade. The agricultural colonization of the west and southwest by Mossi migrants has been the main threat to Burkina's relative ethnic harmony. This vast domestic population movement began around the drought years of the early 1970s as a result of the failure of the Mossi plateau to feed its growing population. The western Bobo and Bwa region, which was underpopulated and significantly more fertile, soon became the main destination of impoverished or destitute Mossi households.[34] Mossi settlements grew increasingly numerous in the west, either at the border of existing villages or eventually becoming villages themselves. Because of the fluidity of land tenancy and property rights throughout the country, the migrants have obtained land for cultivation from local chiefs. Overall, however, the movement has reached an amplitude that threatens both the ecological and the population equilibria. The Mossi, whose presence has reportedly grown by almost 20 percent a year,[35] have deforested large areas for cultivation. With land availability fast shrinking and Mossi population steadily expanding, the local Bobo and Bwa have reportedly grown leery of Mossi settlers. The abundance of rain and land in the 1980s has prevented the emergence of a true crisis. But the elements are present, and a resurgence of drought could be the spark that triggers Burkina's first ethnic explosion.

# Religion

More than 50 percent of Burkina's population is animist. In 1986, the last year for which such data are available, 2,514,261 Burkinabè (about 31 percent of total population) were Muslim and 106,467 (about 1.3 percent) Protestant. In 1990 there were an estimated 826,400 Catholics (about 9 percent).[36] Different beliefs are common within the same ethnic group, and a climate of religious peace prevails among Burkina's faiths.

Animism is most prevalent in the east, among the Gurmanche, and in the southwest around Gaoua. It also dominates—but to a lesser extent—the central Mossi plateau and the Banfora-Bobo region. Islam prevails (at more than 80 percent) among the Peuls, Songhais, and Tuaregs of the north and is still widespread around Ouahigouya and Dedougou. Catholics are nowhere a majority. Their highest concentration (between 20 and 25 percent of the population) is in Ouagadougou and the regions around Koupela (where the first mission was established), Tenkodogo, and Diebougou.[37]

## *Animism*

Animism, the "traditional" religion, which the Mossi call *wend' pous neba*[38] consists of a diversity of beliefs with many local variants. It is still widespread in Burkina and is shared by members of virtually all ethnic groups. Animism is essentially the faith in the enduring existence of ancestors. In Burkina it is also accompanied by the recognition of gods, first among whom is a supreme and distant god the Mossi call Ouennam. Next comes Tenga, the goddess of the earth whose priests are chosen among the *tengbiise*, the children of the land. Animism is thus a system that merges religious with social and economic roles. The *tengbiise* priests, in addition to their custody of the land, are also responsible for relations between the living and the dead. Sacrifices figure predominantly among the animistic rituals. They are made to support prayers for rain, fertility, and other wants.

Although Mossi chiefs converted to imported religions over time, Dim Delobson relativized the meaning of such conversions by stressing how the "Mogho Naba is first of all animist.... [He] can, when necessary, be Catholic or Muslim while respecting the tradition of the ancestors."[39] The *mogho naba* will go to church for Christmas and celebrate the Muslim holy days of Ramadan and Tabaski, but in the end, because of his ceremonial and ritual functions, he must remain an animist.

## *The Penetration of Islam*

Although anterior to Christianity in Africa, Islam is also an imported religion in Burkina. Despite the failed attempt of the Muslim Songhai to impose their faith on the Mossi in a holy war in the fifteenth century, Islam slowly spread in the Mossi plateau between the sixteenth and nineteenth centuries, transmitted by Arab traders and affecting above all merchant communities of Mande origin lo-

*The mosque of Safane. Courtesy of the Centre National de la Recherche Scientifique et Technique.*

cated on the gold, kola and slave routes. These Mande populations, referred to as Dioula by the Malinke of High Niger and Yarse by the Mossi, in turn spread Islam along the regional trade routes.[40]

The Yarse settled in the Mossi country as merchants or marabouts (pilgrims who had returned from Mecca and served as both mystics and schoolmasters) and abandoned their Mande language to speak Moré. By the late eighteenth century, they had penetrated the Mossi hierarchy: Naba Dulugu, who reigned from 1796 to 1825, was converted to Islam by a Yarse he had taken as his imam.[41] Naba Dulugu's conversion was representative of a pattern by which Islam continued to spread in Burkina thanks to the patronage of Mossi chiefs. Indeed, before him, Naba Kom I (1784–1791) is reported to have introduced circumcision and excision among the Mossi and to have allowed the Yarse to stay in villages of his kingdom.[42] Naba Dulugu also allowed the opening of mosques in Ouagadougou. Yet he did not impose the religion onto his people and even forced his son, who looked too favorably upon Islam, into exile. The son returned to overthrow his father, however, and reigned from 1824 to 1842, "an active and convinced believer"[43] who nevertheless continued to fulfill his traditional religious obligations. His own son, the future Naba Kutu (1854–1871), received a Qur'anic education and transferred his traditional religious functions to his ministers.

Despite the progressive conversion of chiefs, Islam had not yet significantly penetrated the animist population as of the nineteenth century. A number of factors were responsible for this situation, according to Jean Audouin and Raymond Deniel: the chiefs' prevention of proselytism due to their fear that Islam would "liberate" the people from "traditional" beliefs by undermining the chiefs' authority; their customary tolerance; their attempt to reconcile Muslim practices with their ancestral culture and rituals; the weakness of Muslim structures; and the resistance of traditional sectors.[44] In consequence, in the words of Michel Izard, "the new religion of the *mogho naba* . . . [would] not become the religion of their kingdom"[45] because of the chiefs' official function in traditional society, where the spiritual and temporal powers are closely linked.

Ironically, the spread of Islam benefited somewhat from the attitude of the French administration in different phases. From 1895 to about 1911, the French officially sympathized with Islam while controlling its spread and influence. By encouraging some sects and allowing the Dioula freedom to travel, on the one hand, but strictly supervising Qur'anic schools, on the other hand, the French administration underhandedly tried to steer Islam toward an "enlightened" version favorable to French values.

Islam was in fact more prevalent in many other French colonies than in Upper Volta, and the French co-optation strategy was inspired by the Muslim factor in its other colonies and by its fear of Islam as a worldwide political force. But French policy remained somewhat favorable to Islam in its religious dimension. France wanted Islam to be genuinely African in its colonies so as to cut it off from the political dimensions of the global Muslim community.[46] But by concentrating its opposition to Islam against marabouts, whom it perceived as more radical in the overall West African picture, France neglected the Dioula merchants, who became the real vector of Islam in Upper Volta.

France's Muslim policy hardened after 1911 as Islam increasingly conflicted with French ambitions to assimilate Africans into French culture. Perceiving no such threat in animism, the colonial administration proceeded to encourage the practice of ancestral customs by the Bambara, Malinke, Bobo, and Mossi and taught customary law in its schools.[47] Meanwhile, the French governor-general stepped up surveillance of Islam in 1911 with the creation of a "Muslim police" force made up of animists and "good" Muslims who overlooked the activities of marabouts. France's paramount social objective in West Africa at the time was to "dissociate black Islam from its Arab roots," to dissociate its religious dimension from its political and social one, and to maintain ethnic and administrative divisions between the different Muslim groups so as to prevent large-scale organization.[48]

In Upper Volta itself Islam was not much of a threat. In 1914 the *cercle* of Bobo-Dioulasso reportedly counted only 2.3 percent Muslims, mostly Dioula. The *cercle* of Ouahigouya was 12.9 percent Muslim, most of them Foulbe-Rimaibe and Yarse-Marka; only 1.2 percent of the *cercle*'s Mossi were Muslim.[49] Yet at the time, Islam was still spreading, slowly and steadily, because of the humble and innocuous Dioula. The administration unwittingly made the spread of Islam even easier, as French-imposed public order allowed the Muslim propagandists to operate without fear of violent reaction.

## Contemporary Influence of Islam

During the longest portion of the colonial period, after World War I, Islam continued to develop as a religion but became increasingly marginalized from politics. Not only was its political potential kept in check by the French, but its own social and political development was hampered by the weakness of Qur'anic schools vis-à-vis their Catholic counterparts, the mission schools.

With independence and the rise to power of Yaméogo, who was adamantly pro-Catholic, the Muslim community felt the need to organize and defend its interests within the new national environment. In 1962 it set up the Communauté Musulmane de Haute-Volta (CMHV) to that effect. But the state, in its undying quest for hegemony, managed to extend its control to the CMHV. State permission became required for construction of mosques, public celebration of holidays, travel to Arab states, and the convening of congresses. Furthermore, the CMHV's honorary presidency fell to the head of state.[50]

After Lamizana, a Muslim, took power in 1966, the political situation of Muslims improved. They gained from official diplomacy when the Arab Muslim world began to look favorably upon Upper Volta. In addition, the break with Israel in 1973 resulted in considerable inflows of Arab aid. Nonetheless, the CMHV was apparently unable to take advantage of the newfound importance of Islam. Not only did it fail to translate the situation into increased political power for Islam, but it also ended up plagued by financial scandals linked to its sudden wealth.[51]

The Islamic movement has not yet recovered politically from this situation and has remained troubled by the factionalism that exists behind the apparent unity of the CMHV. Comprising several different sects, including Qadiriyya (the oldest and largest), Tijaniyya, Hamallism, and Ahamdiyya, the Islamic movement is also divided between traditionalist marabouts and imams on the one hand and Westernized reformists on the other. Furthermore, the CMHV's claim to represent the Muslim community has been challenged since the creation in the mid-1970s of the Association des Sunnites de Haute-Volta, representing a Wahhabiyya current opposed to both the marabouts and the Westernized brand of reformists.[52]

Yet although Muslims are poorly represented in politics and public administration because of their educational shortcomings, they control substantial sectors of the economy, mostly commerce, transport, and construction, and have established profitable relationships with the state, thereby making up in clientelism for some of their lack of direct influence.

## Christianity: Evangelization of the Mossi

The first Catholic missionary to set foot in Upper Volta after the French conquest was Monsignor Hacquard, who toured the Mossi in 1899. A year later the first Catholic mission was set up at Koupela by the Pères Blancs (the Order of White Fathers), followed in 1901 by one in Ouagadougou. The same year Diban Simon

Alfred ki-Zerbo, the father of politician Joseph ki-Zerbo, became the first Voltaic to be baptized a Catholic.[53] Père Blanc Joanny Thévenoud arrived in Upper Volta in 1903. He was to become Ouagadougou's apostolic vicar and bishop from 1921 to 1949. And in 1910 the Soeurs Blanches, or White Sisters, began setting up nunneries around the country.

It is thus apparent that despite the independence of the church from the state in France as in colonial Africa, Catholicism nevertheless arrived in Burkina (as in other French colonies) in the wake of colonization, and both were to be forever associated in African minds. Witness the appellation of whites throughout French West Africa as *nasara*, which comes from the name for the first Catholic missionaries, called Nazareans, disciples of Jesus of Nazareth.[54] Given this identification of church and colonialism, Skinner's contention that "fear was certainly one of the reasons why ... people ... embraced Catholicism" comes as no surprise.[55] Yet there were also positive inducements to conversion. For example, because it was carried out in French, Catholic education provided the basis for clerical employment in the colonial administration and upward social mobility. This dimension of Catholic education became so successful that the missionaries eventually abandoned the teaching of French in order to keep their converts in the rural areas and ensure that they would carry out the proselytizing functions for which they were taught, instead of heading for Ouagadougou to seek employment in the administration.[56] Among those who stayed in the villages, many became catechists and some priests. The first three Catholic priests were ordained in 1942. Because evangelization had started so early in Burkina and because the populations were somewhat more receptive than in more Islamic countries, the country became a seedbed for West African Catholic dignitaries. Dieudonné Yougbaré became the first non–French West African Catholic bishop in 1956, and Paul Zoungrana, who was made an archbishop in 1960, later became West Africa's first cardinal.[57]

Yet Catholicism grew slowly among Voltaics. The biggest obstacle to its growth seems to have been the prohibition of polygamy, which was contradictory both to tradition and to the Voltaics' desire to maximize their chances of having children. In addition, because of the custom that Mossi women who have given birth return to stay with their parents until the child is weaned, after about two years, monogamy was unpopular with men, who feared having no one to share the household chores with or to satisfy their sexual needs for a prolonged period of time.[58] From this point of view, Islam, which tolerated up to four wives, was much more practical. Individual conversions to Catholicism were also made difficult, for if converts were ostracized by their extended families, they would be at pains to cultivate their fields without the help of others. People found ways to get around the rigors of Catholicism, however. Men would take mistresses with or without their wives' approval. And men who had converted in order to marry Catholic women educated by nuns often moved to other villages with their wives so they could adopt non-Catholic lifestyles.

*President Compaoré welcomes Pope John Paul II to Ouagadougou. Courtesy of* Sidwaya.

Thus, overall, Catholicism progressed slowly during the colonial period in terms of number of believers. But at the same time it gave its converts the formal and modern education that would be a powerful tool of advancement under both the colonial and the independent states, thereby empowering the Catholic minority well beyond its arithmetic importance.

Protestant American Assemblies of God also opened missions in Upper Volta, beginning in Ouagadougou, Kaya, and Yako in 1926, but both their size and influence have since remained insignificant.

## Contemporary Influence of Christianity

Because of its educational functions, the Catholic Church since independence has had a political role disproportionate to the number of its adherents in the country. For a long time, the only secondary school in Upper Volta was the seminary where future clergy and politicians befriended one another. Until the nationalization of education in 1969, Catholic schools provided education to one-third of all children in primary and secondary schools.[59] Today the link between church and politics has loosened considerably. Nevertheless, Catholics remain overwhelmingly overrepresented within the political class.

Independence began with a near symbiosis of state and church, as Yaméogo strongly favored the latter's interests. The church paid him back in unconditional support until he divorced and remarried in 1965. The Catholic Church was more subdued under Lamizana, although the latter was forced to recruit mostly Catholic ministers for his government for lack of properly trained Muslim alternatives. Nevertheless, when Lamizana was overthrown in 1980, the church must have been relieved, for Archbishop Zoungrana publicly commented that the coup was a blessing. The 1980s would see a growing schism between church and state and increasing intimidation of the former by the latter. The church and the CNR government were hostile toward each other. It appears that Sankara considered the church an instrument of oppression and thus a worthy target for his revolution, although he himself was a Catholic and apparently encouraged his wife to be baptized. Blaise Compaoré, also a Catholic, has resumed more cordial relations between the state and the church.

The pro-Arab policies of Lamizana and the public welcome of his overthrow by Catholic authorities triggered some tensions with the Muslim community. Despite these frictions and the rise of militant Islam in other parts of Africa, interfaith relations remain surprisingly good in Burkina. From an anecdotal point of view, a mixed couple (she a Catholic, he a Muslim) I have befriended over the course of my visits to Ouagadougou since the early 1980s is not shocking to other Burkinabè, nor are the pair in the least marginalized.

## Sick and Illiterate: The Status of Health and Education

Burkina ranks at the bottom in terms of social indicators of development with respect to both health and education. Progress in health has been slow, as authorities have had to fight several endemic conditions, and education, which has often fallen victim to politics, has been expensive and unproductive.

### The Precariousness of Life

In 1992 infant and child mortality rates were estimated at 132 and 196 per 1,000.[60] While this marked progress since independence in 1960, when infant mortality was 263, the rate of improvement has slowed significantly since the early 1980s and the infant and child mortality rates are well above the average of 91 and 146 for low-income countries. Burkina is not alone in this predicament, however. Several sub-Saharan African countries experienced a slowdown (and some a decline) in social indicators in the 1980s.[61] Similarly, Burkinabè's life expectancy, which stood at thirty-seven years in 1960, was still estimated at around only forty-eight years in 1992,[62] while the average for low-income countries (excluding China and India) was fifty-six.

The primary causes of poor health and morbidity in the adult population are malaria and gastrointestinal, infectious, and parasitic diseases (such as onchocerciasis or river blindness). AIDS also made major headway into Burkina in the 1980s. Among children the killers are measles, meningitis, and malnutrition (some 45 percent of children under five were estimated to be malnourished as of 1992).

The health situation is aggravated by the overall inadequate sanitation. The current World Bank estimate that 67 percent of the population has access to safe water seems significantly exaggerated, as only 25 percent had such access in the mid-1970s.[63] There is one physician per 57,000 people, one nurse per 1,682, and one hospital bed per 3,392. In addition, these figures hide substantial differences between the relatively better access to health care of urban dwellers over their rural counterparts. Despite a major vaccination campaign under the CNR in the 1980s, only 40 percent of the children are immunized against measles and 30 percent against diphtheria, poliomyelitis, and tetanus (DPT).[64]

While the battle against malaria has suffered a considerable setback throughout Africa with the development of strains more resistant to medication, Burkina has recorded at least one major health success with the virtual elimination of onchocerciasis. Meanwhile, however, AIDS has become a growing scourge in Burkina.

Onchocerciasis, a parasitic disease carried by blackflies, until recently affected some 10 million people throughout West Africa. It causes blindness and severe skin irritation and led to the desertion of several villages and land in the river basins of Burkina and elsewhere. Since its creation in 1974, the Onchocerciasis Control Program (OCP), a branch of the World Health Organization, has conducted aerial spraying of blackfly breeding sites, which as of 1984 had interrupted the transmission of the disease in more than 90 percent of the OCP area throughout West Africa.[65] In Burkina alone (which hosts the OCP headquarters and where about half the program area lies) it is estimated that the OCP prevented 27,000 cases of river blindness in its first eight years and that 16 percent of previously unused land had been settled during the program's first ten years.[66]

Yet the gains from the onchocerciasis campaign are of little comfort in light of the spread of AIDS. As of mid-1993 there were 1,263 officially reported cases of deaths from AIDS in Burkina.[67] Although this is not much by absolute standards compared to some other African countries, it is quite large for West Africa and for Burkina's modest size. Migration to Côte d'Ivoire and Ghana is no doubt in part responsible for this rapid spread of AIDS in Burkina. Côte d'Ivoire has been found to be the West African country with the highest incidence of AIDS, especially in Abidjan and the coastal areas, where many Burkinabè migrants live. Until recently, however, public authorities still showed much misunderstanding of and prejudice toward the disease. In a 1987 speech to the second national conference of the CDRs, Sankara added insult to injury for AIDS sufferers, saying that the "poor do not have AIDS. [Only] white people, rich people, and exploiters have this disease. . . . Every African who has AIDS is a colonized African."[68]

## The Privilege of Education

The mission schools of the Pères Blancs and Soeurs Blanches were the first centers of formal education to be established in Upper Volta early in the twentieth century. They taught religion and the French language to the first generation of

Voltaic graduates. Once the colonial administration was in place, however, public schools were established. In July 1920 the first public primary and vocational school opened in Ouagadougou. Its aim was to prepare candidates for the Ecole William Ponty, an elite colonial school in Dakar that trained future colonial administrators. In 1920, too, the first official girls' school was created in Ouagadougou. Burkina's formal educational system currently requires six years of primary and seven years of secondary schooling in French and is officially free, although the actual expenses incurred in sending a child to school remain above the means of most households, especially in rural areas.

Partly as a result of the paucity of means of both parents and the state, the education system falls far short of meeting the needs of Burkina's population. As of 1990, 91 percent of women and 82 percent of men were still illiterate.[69] Of those who know how to read and write, about 75 percent are trained in French.[70] The literacy rate is not expected to improve dramatically over the coming years, as a mere 30 percent of children aged six to eleven were in primary school in 1991 (as against 79 percent for low-income countries in general). This was up from 8 percent in 1960 and 13 percent in 1970, but marked a decrease from 32 percent in 1985. Yet despite the percentage drop in enrollment, classes remain overcrowded. The average primary school student-teacher ratio went from 44:1 in 1970 to 58:1 in 1991. About 10 percent of the children in primary school are enrolled in private establishments. This number would have been greater had the church not returned all its education functions to the state in 1969 following two years during which the state refused to raise its subsidies to the Catholic education system. The state took over the Catholic schools in the fall of 1969 and replaced clerical teachers with civil servants.[71]

Enrollment in secondary school amounts to 8 percent of the children aged twelve to eighteen, up from virtually none in 1960 and 1 percent in 1970. Once again Burkina's performance is below other low-income countries, which enroll on average 28 percent of the secondary-school-age cohort.[72] The role of private institutions is paramount at this level of education, as they enroll no less than about 50 percent of all children in secondary school.[73] About 1 percent of those above eighteen years of age register in college, some 5,400 of them at the University of Ouagadougou (which counts 390 professors)[74] and a few others abroad, mostly in Senegal and in France.

Between 10 and 15 percent of government expenditure is usually allocated to education and for the most part covers teachers' salaries and academic scholarships. Until 1989 every university student received a state scholarship provided he or she held a baccalaureate (high school completion certificate) and was not older than twenty-two when entering university. There was no income criterion. In order to retain the scholarship over the whole course of study, students could not fail more than once per cycle of two years, that is, they could repeat a year every other year. The scholarships amounted to CFAFr 37,500 per month from the first to fourth year, then CFAFr 48,500 for those in longer programs, such as medicine. Following a decision inspired by the CDRs in 1986, a less liberal system was adopted in 1989. A given quota of scholarships based on merit have since been awarded to students in decreasing order of their performance on the baccalaureate exam, still without consideration for family income. First-year students receive CFAFr 25,000 per month. An increase of CFAFr 5,000 is granted every following year until the fourth year. Students in longer programs receive CFAFr 48,500 from their fifth year onward. Students who do not receive a scholarship are still admitted to university. Until 1990–1991 they received a "government aid" package that was later discontinued. This still highly favorable policy toward tertiary education indicates that university students may have been privileged at the expense of primary school students.

Given the failure of the primary school system to provide universal education, Burkina's successive governments have occasionally encouraged alternative and informal education as well as community involvement in building schools and classrooms. One such project was the setting up of rural education centers in the early 1960s. Their objective was to provide a low-cost alternative to primary school by training unschooled children of farmers in reading, writing, math, and agricultural techniques over a three-year course. The government had to end the program in 1973 with only 737 centers and some 24,000 enrolled pupils, one-sixth of the original target. One reason these centers remained unpopular is not only that most of the classroom time was allocated to learning the language of instruction, French, but they also did not award diplomas and provided little hope of social advancement. They were replaced by young farmers' training centers run by the Ministry of Agriculture. But these, too, proved inefficient and costly and were phased out in the 1980s.[75]

Qur'anic and Franco-Arab schools (*medersa*) also provide Muslim children and young adults with a private alternative to official schooling. They have taught the 22 percent of nationals who are literate in Arabic.[76]

Education has also often fallen victim to politics. Numerous teacher strikes have been milestones in the country's political history. It was a fifty-three-day strike by grade school teachers that led to the November 1980 coup that overthrew Lamizana. In 1984 Sankara dismissed hundreds of unionized primary school teachers when they protested his policies. They were replaced by young, untrained "volunteers" whose credentials as revolutionaries were far superior to their credentials as teachers.

Politicization occurs among students as well. Demonstrations of high school students were often instigated or used by the CNR to provide popular support for specific measures or policies. Moreover, the University of Ouagadougou and its student unions have been the birthplace of most of Burkina's radical political organizations. In a sign that they remain politically active and a serious concern to the state, students were still arrested and allegedly tortured by Compaoré's police as recently as 1991.

## Women and the "Other Half of the Sky"[77]

In his preface to Dim Delobson's study of the Mossi published in 1932, the French colonial administrator Robert Randau dramatically described the predicament of Mossi women:

> The principal obstacle our civilizing action is faced with is not the mystical beliefs, the miscellaneous taboos, the customs bequeathed by the ancestors. It is the slavery in which the black holds his wife. One cannot even talk of polygamy among the Mossi, but of human beings treated like beasts of burden and valued less than a horse or a cow.[78]

Written as they were at a time when equality of the sexes was far from established even in France, these words indicate the trying odds the Mossi women faced. It is no wonder, then, that at the end of the twentieth century the plight of Burkina's women is still overwhelmingly difficult.

This section first reviews the customary role of women in society and in the economy. It then goes on to assess the changes in women's status that the 1983 revolution attempted to bring about.

### *Women in Society*

Women have played a central role in the production and reproduction of the Mossi system and of its social order. It should be recalled that the Mossi kingdoms owe their origin to a woman, Niennega, who left her father's country to found a dynasty in Yanga. It should also be remembered that it was through marriage outside of her lineage, with Ryallé, that she established these foundations. The institution of marriage has indeed been the means by which women have customarily realized their historical role in Mossi society. And as if to emphasize that their historical significance has nothing to do with their individual worth as women, the women credited with the most importance reportedly had the features of men. When Ryallé pledged allegiance to Niennega, he did so thinking she was a man. And later, their son Ouédraogo founded Tenkodogo by marrying a local Ninissi woman, Pougtoenga, who allegedly had a beard.[79]

Skinner asserts that Mossi "women emerged as political actors primarily because of their statuses and roles *within* kinship groups" and that these roles included those first of "state-building warriors" and then state consolidators, "since their marriages often linked the conquerors and the conquered."[80] This last point is paramount to the foundation of Mossi society, which resulted from the alliance, sealed in marriage, of the conquering Dagomba and the allegedly happily conquered Ninissi. Marriage and assimilation have indeed been the instruments of Mossi expansion and hegemony over the centuries. Mossi marriage is thus exogamous by nature, originally taking place between the Dagomba and the Ninissi and later between different lineages of the Mossi.

But even if women are essential to the system, that does not imply that they control their destiny. On the contrary, customary Mossi marriages frequently occurred in the form of an exchange of women between male-dominated lineages. The institutionalization of this exchange is called *pogsyure*.[81] According to *pogsyure*, a man from one lineage gives his daughter in marriage to a man of another lineage in exchange for the first daughter to come out of this marriage, who will then become part of the former household and lineage. *Pogsyure* is thus a mechanism of circulation and exchange of women.[82] It usually operates between two different lineages and commits all the members of the lineage.[83] The lineage that gives a woman is entitled to a woman from the other lineage even if no child comes out of the first union.

The context in which *pogsyure* has developed on a large scale is that of exchange involving the chiefs, the *nanamse*, in which case it is called *napogsyure*. A *naba* gives a woman to a man in exchange for a service (or years of service). This man thereby commits to give the *naba* the firstborn of his union with that woman. If it is a boy, he becomes a servant at the court; if it is a girl, she will join the reserve of women awaiting to be given away in *napogsyure*.[84] Kings are thus at the center of a major system of exchanging women. This exemplifies how women, who hold the power to reproduce, are managed like a scarce resource, which fits Hans Binswanger and John McIntire's assessment that in semiarid tropical regions where land is abundant and population density low, property will take the form of people rather than land.[85] Although the authors had slavery in mind, *pogsyure* is somewhat akin to it, women being exchanged much like a commodity. Furthermore, in the case of *napogsyure* it is equivalent to the "capitalization of women as a means of power,"[86] as chiefs receive more women than they give away. For Izard,

> the accumulation of women or of servants ... is an accumulation of forces of production.... The role of polygyny in the capacity to valorize land resources is too evident to be insisted upon.... The accumulation of women [by kings] is on the one hand the accumulation of an immediate labor force, of exogenous origin, and on the other hand, to the extent that women give birth to children, of a future labor force, of endogenous origin, and of matrimonial exchange value.[87]

And he concludes that *pogsyure* was "a fundamental element of the implementation of economic accumulation and of the establishment of relations of political dependency."[88] Beyond this accumulation function, however, *pogsyure* also contributed to the integration of different lineages, the creation of ethnic identity, and the reproduction and growth of the social system.

Not surprisingly, *pogsyure* met with some opposition during colonial times. The status of women was one of governor Hessling's main concerns. In a circular on "the condition of the native woman," he complained that the Mossi woman was "so often subjected to a special form of slavery which, notably at the time of marriage, reduces her to the role of a domestic animal," and he vowed to change these traditions.[89]

The Catholic Church, too, did much to reduce the importance of *pogsyure*, though not always knowingly. The Catholic missions' recruitment of girls upset many chiefs, as it removed them from the circulation network. In addition, Mossi men could not marry these Catholic women without converting, too, which led to a loss of the traditional function of marriage. It was then not uncommon for a Mossi man to run away with his bride and return her to traditional social modes.

The influence of state, religion, and modernization, together with changing demographics, have gradually eroded the *pogsyure* system, which seems no longer prevalent. The concurrent practice of polygamy lingers, however. Two types of marriage are currently acknowledged. The customary marriage may be either polygamous or monogamous, while the formal "modern" marriage is monogamous only. According to the 1985 census, 50 percent of men aged twelve or up are married and 18 percent are polygamous. The proportion of polygamous men increases with age to peak at sixty, suggesting that polygamy is mostly a function of wealth. Married men have on average 1.6 wives (64 percent of married men are monogamous, 25 percent have two wives, 7 percent have three wives, and 3 percent have four or more). In 1975 only 15 percent of men aged twelve or above were polygamous, but married men already averaged 1.5 wives in 1960. From the reverse perspective, 57 percent of wedded women are married to a polygamous husband.[90] Polygamy is also more widespread in rural than urban areas, betraying the agricultural labor-power function of spouses.

Another institution that remains prevalent in Burkina and that has contributed equally if not more than *pogsyure* to women's submission is the practice of excision. Excision is the surgical removal of all or part of a woman's clitoris and may also include the ablation of the minor labia and part of the major labia. It results in frigidity and is widely practiced in Burkina by most ethnic groups. Almost all women are excised, according to some sources.[91] One of the main rationales for excision is to prevent women's sexual promiscuity. Another frequently cited reason is the belief that if a child touches his or her mother's clitoris upon being born, he or she will soon die.[92] This appears to be a symbolic ban on incest.

Excision is performed by older women on young girls and sometimes on young women. In any case, it precedes marriage, as a nonexcised woman will be hard put to find a husband, at least outside the main urban areas and among the less-educated people. The excision is often performed simply with nonsterilized scissors or knives and obviously without anesthesia. Although several women immobilize the girl being excised to prevent accidents, there are frequent instances of death and serious injuries including infections, tetanus, and scarring that may later result in painful intercourse.

As earlier mentioned, the labor of women is essential, and it is in rural areas, where the harsh demands of agricultural work add to their low social status, that their predicament is worse. Rural women frequently walk 3 to 6 miles daily, before dawn, to fetch wood and water. They prepare the day's meals in the early morning and spend most of the day working in the collective fields. Upon returning to the compound, they tend to their own personal plots, make flour, cook for the household and look after the children and their husband's needs. It is estimated that altogether women are responsible for 60 to 80 percent of all agricultural work.[93]

In urban areas women's economic role is substantial in markets and in the informal sector, where they may manage to reach some level of economic independence.[94] They sell fruits and vegetables that they or other women have cultivated, eggs, spices, and fish. They exclusively make *dolo,* the popular sorghum beer; *beignets,* and pottery and spin cotton.[95] They are, however, underrepresented in the formal economy and public administration, where they make up less than 10 percent of the salaried private sector and less than 20 percent of the public service; they also tend to be limited to middle-management positions. But women in the formal economy, particularly civil servants, are substantially better off than their rural and formally unemployed counterparts. The Labor Laws Act of 1962 allows female public servants fourteen weeks of paid maternity leave and one hour off per day for breast-feeding for fifteen months thereafter.[96]

## *Women in the Revolution*

Given the status of women, it comes as no surprise that a self-proclaimed revolutionary regime made the improvement of their lot one of its battle cries. On several occasions in public speeches, Sankara stressed the joint nature of his revolution and women's liberation,[97] and his regime did attempt a great deal. The Sankara government launched an education campaign against excision, which was to be followed by its criminalization. While virtually no legislation aimed at radically altering the status of women had been introduced by governments since independence, a ministry of family development was created under Sankara to "coordinate measures aimed at improving the situation of women and families" and a family law plan was devised.[98] The "upgrading of the status of the woman as an agent for development" was made the fourth objective in the development strategy of the country's 1986–1990 five-year development plan.[99] A National Week of the Woman was held in 1985. More women joined the government, and they received somewhat more significant portfolios. In November 1988 the Individual Persons and Family Law Act, initiated under Sankara, was passed. Among other things, it established monogamy as the basis of common law marriage—yet, oddly retained "optional polygamy"—and asserted the principle of free choice of spouse.

There is a gap, however, between laws and reality, and the legal reformism of the CNR fell short of closing it. Having what is considered the most advanced family law in the Sahel has been of little use in changing the condition of women in rural areas. Despite its official abolition by the CNR in the 1985 legal reform, customary law has continued to rule over many matters of individual relationships in villages, where the legal innovations of the CNR were met with hostility by a rural population that had already received evidence that the government was otherwise intent on destroying its social structures. The campaign against excision met with serious resistance. Social indicators of development remained biased in men's favor. For example, as recently as 1991, there were only sixty-two girls in primary school and fifty in secondary school for every 100 boys.[100] Finally, women's access to land was not improved. On the contrary, the land reform tended to deprive them of land and to downgrade "their socio-economic status in rural environments," according to a study released by the OECD's Sahel Club.[101]

Meanwhile, as the CNR policies effectively failed to emancipate women, so in its aspiring totalitarianism the CNR made sure to keep women within the well-defined bounds of the revolution and within the grip of the state. To this end the CNR created the Union of Burkinabè Women (UFB) in September 1985 as an organization for the mobilization of women for the revolution. The revolution brought more hope than concrete results with respect to women's plight.[102] To be fair, in the short run the state can probably do little, as its reach and its "modernism" do not necessarily extend to segments of society such as that of rural women and the institutions of their environment. It is possible that beyond the local and specific initiatives of some nongovernment organizations (NGOs), little else can be done at the national level before economic development further changes the structure of social and sexual relations.

## The Culture of the State

Sixty ethnic groups, three language families, three core religions, many diverse schooling experiences, and highly differentiated gender roles and relations: social and cultural pluralism virtually defines Burkina. Yet although the country is a collection of cleavages, it has had uninterrupted social peace since independence.[103] Possible reasons for this were addressed earlier in this chapter. They included the assimilationist outlook of the Mossi; the clear-cut split between state and civil society, which has prevented any group from "owning" the state and thus has deflated reasons for antagonism; the succession of Catholic and Muslim leaders as heads of state; and the country's lack of resources, which gives little stake to social competition.

Beyond its particularisms, however, contemporary Burkina also harbors a single modern popular culture, spread across the country—with the help of the state—through 225,000 radio receivers and 45,000 television sets.[104] But one of its most successful vehicles of expression has been the motion picture, a genre in which Burkina has developed a remarkable advantage. Every other year, the state sponsors the highly popular Festival Panafricain du Cinéma de Ouagadougou, which attracts crowds of cinephiles from well outside Africa.[105] It has earned Ouagadougou the nicknames of "the Hollywood" and "the Cannes of Africa." The

*Samo dancers decorated with cowries in Kiembara. Courtesy of the Centre National de la Recherche Scientifique et Technique.*

country has movie studios and a few internationally renowned producers, including Gaston Kaboré who in the early 1980s was responsible for *Weend Kuni*, the celebrated account of rural life in Burkina seen through the eyes of children, and Idrissa Ouédraogo, who produced *Tilai, Yaaba,* and most recently *Samba Traoré*, the story of a burglar who returns to his village around Banfora with his ill-acquired wealth only to have remorse, suspicion, and justice eventually catch up with him.

Every year in which FESPACO does not take place, the state sponsors National Culture Weeks, festivals of cultural presentations such as dance, theater, and singing, held in a decentralized and different location each time.

In Burkina modern culture does not deny ethnic, regional, or religious particularisms, as happens in other parts of Africa, where individual identification at any level below national is reproved. This is so because of the historical lack of cultural threat to the integrity of the state in Burkina. Nevertheless, the national expression of culture represents specific and regional cultures as local folklore, not as alternative systems of legitimation. In other words, culture is not generally allowed to be a source of political challenge. Its political component is virtually

Society and Culture

*A still from* Weend Kuni, *a film by Gaston Kaboré. Courtesy of Gaston Kaboré.*

banned, unless it can be embraced by the state. These are the constraints modern African statehood usually imposes on its cultures. They are inherent in the nature of the African state and date back to independence, when African leaders chose a European-inspired political culture according to which, in Basil Davidson's words, "a successful nation-statism in Africa must dispense with, or better still ignore, every experience of the past. Tradition in Africa must be seen as synonymous with stagnation. The ballast of past centuries must be jettisoned as containing nothing of value to the present."[106]

Although somewhat overstated, the point is nevertheless well taken. Burkina's statehood, like that of other African countries, is culturally European. Even Sankara's cultural policy never resolved its own contradictions, stressing on the one hand the authenticity of local cultures against "imperialism" and accordingly renaming the country and, on the other hand, imposing its European-designed state and party system to the smallest corners of society and clashing head-on with local traditions and institutions stigmatized as "feudal lords" or as "the decadent values of our traditional culture."[107] Although no such extreme was reached before or after Sankara, his was only the most radical expression of a contradiction all regimes live with and reproduce, that between the ostensibly rational-legal culture of the state and the traditional cultures of society, to use a typology Max Weber originated in the context of regime legitimacy but that applies as well to the context of culture.

## Notes

1. World Bank, *World Development Reports (WDR)* for 1980 and 1994 (Oxford: Oxford University Press, 1980 and 1994).

2. Ibid.

3. All the following numbers are from the Institut National de la Statistique et de la Démographie (INSD), *Recensement général de la population Burkina Faso 1985.* Analyse des résultats définitifs (Ouagadougou: Imprimerie de l'INSD, 1990).

4. These figures are gross estimates based on the 1985 census figures.

5. Institut National de la Statistique, *Deuxième recensement général de la population du 10 au 20 décembre 1985: Principales données définitives* (Ouagadougou: Ministère du Plan et de la Coopération, n.d.).

6. INSD, *Burkina Faso: Recensement,* 152.

7. Ibid., 153.

8. This information is inferred from the 1985 census. The census, however, does not mention ethnic groups but spoken languages, which I use here as a proxy for ethnic identity. To be more precise, then, I would say that 50.2 percent of Burkinabè speak Moré.

9. *Enquête démographique par sondage en République de Haute-Volta, 1960–61,* cited in Norbert Nikiéma, "La Situation linguistique en Haute-Volta: Travaux de recherche et d'application sur les langues nationales" (UNESCO, Paris, 1980, mimeographed). I am not aware of more recent ethnic data and have assumed an equal rate of growth for all ethnic groups. It is possible, however, that the Mossi grew faster than some other groups, for their share grew from 48 percent in 1960 to 50.2 percent in 1985 (although, as explained in note 8 above, the latter figure refers to language spoken rather than ethnicity). In any case, the difference does not appear significant enough to alter the overall picture of the population's ethnic distribution.

10. See Institut Géographique du Burkina (IGB), *Burkina Faso: Carte Linguistique.* (Ouagadougou: IGB, 1988). There are, in addition, several small groups that belong to linguistic families more prevalent in other countries, such as the Dogon, the Songhai, and the Haussa.

11. Claudette Savonnet-Guyot, *Etat et sociétés au Burkina: Essai sur le politique africain* (Paris: Karthala, 1986), 19.

12. Ibid., 125.

13. John D. Hargreaves, ed., *France and West Africa: An Anthology of Historical Documents* (London: Macmillan, 1969), 165.

14. I set aside the other possible objection, namely, that the Mossi themselves are not a unified group. The *mogho* is 63,500 sq km and counts several relatively autonomous kingdoms, often engaged in rivalries with one another. Any collective action on their part would be quite hypothetical.

15. Elliott P. Skinner, *The Mossi of Burkina Faso: Chiefs, Politicians and Soldiers* (Prospect Heights, Ill.: Waveland Press, 1989), 171.

16. Ibid., 184–185.

17. Daniel Miles McFarland, *Historical Dictionary of Upper Volta.* Metuchen, N.J.: (Scarecrow Press, 1978), 33.

18. Philippe Lippens, *La République de Haute-Volta* (Paris: Berger-Levrault, 1972), 28.

19. Ibid., 28.

20. Skinner, *The Mossi*, 186.
21. Quoted in ibid., 188.
22. Thomas Sankara was a Silmi-Mossi, a term that refers to individuals born of a Mossi mother and a Peul father.
23. See Chapter 2.
24. Larba Yarga, "Modernisation administrative et autorité traditionelle en Haute-Volta" (Université de Nice, 1975, mimeographed), 60.
25. Lippens, *La République*, 27.
26. René Otayek, "Burkina Faso: Between Feeble State and Total State, the Swing Continues," in Donal B. Cruise O'Brien, John Dunn, and Richard Rathbone, eds., *Contemporary West African States* (Cambridge: Cambridge University Press, 1989), 16.
27. Conseil National de la Révolution, *Discours d'orientation politique (DOP)* (Ouagadougou: Ministère de l'Information, 1983).
28. See Otayek, "Between Feeble," 29.
29. Elliott P. Skinner, "Sankara and the Burkinabé Revolution: Charisma and Power, Local and External Dimensions," *Journal of Modern African Studies*, 26, 3 (1988): 444.
30. As argued as well by Savonnet-Guyot, *Etat*, and Dominique Zahan, "The Mossi Kingdoms," in Daryll Forde and P. M. Kaberry, eds., *West African Kingdoms in the Nineteenth Century* (London: Oxford University Press for the International African Institute, 1967), 152–178.
31. Karl Marx and Friedrich Engels, *The Communist Manifesto* (New York: Monthly Review Press), 1964, 2–3, 10–11.
32. See Pascal Labazée. "Discours et contrôle politique: Les avatars du sankarisme," *Politique Africaine*, 33 (1989): 11–26.
33. Jean-François Bayart, "Civil Society in Africa," in Patrick Chabal, ed., *Political Domination in Africa: Reflections on the Limits of Power* (Cambridge: Cambridge University Press, 1986), 113.
34. Michel Benoît considers this movement to be the continuation of the Mossi's historical expansionism. See *Oiseaux de mil: Les Mossi de Bwamu (Haute-Volta)* (Paris: ORSTOM, 1982).
35. Bernard Tallet, "Espaces ethniques et migrations: Comment gérer le mouvement?" *Politique Africaine*, 20 (December 1985): 69.
36. *Africa South of the Sahara 1993* (London: Europa Publications, 1992), 191.
37. Jean Audouin., "L'Evangélisation des Mossi par les Pères Blancs: Approche sociohistorique." Doctoral dissertation, Ecole des Hautes Etudes en Sciences Sociales, Paris, 1982, 525–526, based on 1973 data from the Ministry of Interior.
38. McFarland, *Historical Dictionary*, 153.
39. A. A. Dim Delobson, *L'Empire du Mogho Naba: Coutumes des Mossi de Haute-Volta* (Paris: Donat-Montchrestien, 1932), 203.
40. Jean Audouin and Raymond Deniel, *L'Islam en Haute-Volta à l'époque coloniale* (Paris: L'Harmattan and INADES, 1978), 12–13. The study of Islam (and religion in general) in Burkina has benefited tremendously from the work of Audouin and Deniel. They have inspired many of the insights in this section.
41. Ibid., 16.
42. Ibid., 17.
43. Ibid., 18.

44. Ibid., 19.

45. Quoted in ibid., 19.

46. Note that during World War I the Ottoman Empire was an ally of Germany and Muslims were thus sometimes perceived as agents of the German enemy.

47. Audouin and Deniel, *L'Islam*, 31.

48. Ibid., 41.

49. Ibid., 34–35, based on surveys by the French administration.

50. René Otayek, "La Crise de la communauté musulmane de Haute-Volta: L'islam voltaïque entre réformisme et tradition, autonomie et subordination," *Cahiers d'Etudes Africaines*, 24, 3 (1984): 315.

51. Ibid., 303–304.

52. Ibid., 310.

53. See Joseph ki-Zerbo, *Alfred Diban, premier chrétien de Haute-Volta* (Paris: Cerf, 1983).

54. Audouin, "L'Evangélisation," 140.

55. Elliott P. Skinner, "Christianity and Islam Among the Mossi," *American Anthropologist*, 60, 6 (December 1958): 1111.

56. Ibid.

57. McFarland, *Historical Dictionary*, 67.

58. Skinner, "Christianity," 1113–1114.

59. Lippens, *La République*, 29.

60. World Bank, *WDR 1994*, 214. Infant mortality is defined as the mortality rate of infants under one year old per 1,000 live births. Child mortality, defined as the mortality rate of children under five years old per 1,000 live births, was 186 for girls and 205 for boys.

61. See Jacques van der Gaag, Elene Makonnen, and Pierre Englebert, *Trends in Social Indicators and Social Sector Financing*, World Bank Working Paper Series 662 (Washington, D.C.: World Bank, 1991).

62. World Bank, *Social Indicators of Development* (Baltimore: Johns Hopkins University Press, 1994), 52.

63. Ibid., 53.

64. Ibid., 52.

65. Stephen D. Younger and Jean-Baptiste Zongo, "West Africa: The Onchocerciasis Control Program," in Economic Development Institute of the World Bank, *Successful Development in Africa: Case Studies of Projects, Programs and Policies* (Washington, D.C.: World Bank, 1989), 27. See also Kodwo Ewusi, S. K. Daapah, and Clement Atriadeke, *The Development and Resettlement of Oncho-freed Zones in the Volta River Basin of West Africa: A Report to the Food and Agriculture Organization* (Legon: Institute of Statistical, Social and Economic Research, University of Ghana, 1985), and David Wigg, *And Then Forgot to Tell Us Why . . . : A Look at the Campaign Against River Blindness in West Africa*. (Washington, D.C.: World Bank, 1993).

66. Younger and Zongo, "West Africa," 34–35.

67. World Health Organization figures quoted in *AIDS*, May-June 1993.

68. Quoted in Labazée, "Discours et contrôle," 25.

69. World Bank, *WDR 1994*, 162.

70. INSD, *Deuxième recensement*, 25.

71. *Africa Contemporary Record*, vol. 2 (New York: Africana Publishing, 1969–1970), B618–619.

72. World Bank, *WDR 1994, WDR 1988,* and *WDR 1980.*
73. Based on 1980–1981 data. World Bank, *Upper Volta: Investment in Human Resources. Country Economic Memorandum* (Washington, D.C.: World Bank, 1983), 70.
74. *The World of Learning 1994* (London: Europa Publications, 1993).
75. Marlaine E. Lockheed, Adriaan M. Verspoor, et al. *Improving Primary Education in Developing Countries* (Oxford: Oxford University Press, 1991), 32.
76. INSD, *Recensement général,* 257–258.
77. In the *DOP* of 2 October 1993, Sankara made the point of sexual equality by saying of women that they "bear upon themselves the other half of the sky" (Conseil National de la Révolution, *DOP,* 36).
78. Robert Randau in Dim Delobson, *L'Empire,* v.
79. Ibid., 5.
80. Skinner, *The Mossi,* 11.
81. Sometimes spelled *pugsyure.*
82. Michel Izard, *Gens du pouvoir, gens de la terre* (Cambridge: Cambridge University Press, 1985), 510–511. See also Dim Delobson, *L'Empire.*
83. Skinner, *The Mossi,* 24.
84. Izard, *Gens,* 514.
85. Hans Binswanger and John McIntire, "Behavioral and Material Determinants of Production Relations in Land-Abundant Tropical Agriculture," *Economic Development and Cultural Change,* 36, 1 (October 1987): 97.
86. Savonnet-Guyot, *Etat,* 117.
87. Izard, *Gens,* 522.
88. Ibid., 524.
89. Quoted in Skinner, *The Mossi,* 169.
90. INSD, *Recensement général.*
91. Gilbert Tarrab and Chris Coëne, *Femmes et pouvoirs au Burkina Faso* (Quebec: G. Vermette and L'Harmattan, 1989), 21.
92. Ibid., 21.
93. Marthe Doka Diarra and Marie Monimart, "Women and Sustained Development in the Sahel: An Analysis of National and Aid Agencies' Policies in Burkina Faso and Mali" (OECD/Club du Sahel, Paris, 1989, mimeographed), 5.
94. See Ellie Bosch, *Les Femmes du marché de Bobo: La vie et le travail des commerçantes dans la ville de Bobo-Dioulasso au Burkina Faso* (Leiden: Rijksuniversiteit te Leiden, 1985).
95. Tarrab and Coëne, *Femmes et pouvoirs,* 79–80.
96. Diarra and Monimart, *Women,* 8.
97. For example, in the *DOP,* 35–37 and "The Revolution Cannot Triumph Without the Emancipation of Women," a speech delivered in Ouagadougou on 8 March 1987 for International Women's Day, reprinted in Thomas Sankara, *Thomas Sankara Speaks: The Burkina Faso Revolution, 1983–1987,* trans. Samantha Anderson (New York: Pathfinder, 1988), 201–227.
98. Diarra and Monimart, *Women,* 7.
99. Ibid., 9.
100. World Bank, *WDR 1994,* 218.
101. Diarra and Monimart, *Women,* 10.
102. For the point of view of hope, see Christine Benabdessadok, "Femmes et révolution, ou comment libérer la moitié de la société," *Politique Africaine,* 20 (December 1985).

54–64. Also of interest from this point of view is the FP government's establishment of a Fonds d'Appui aux Activités Rémunératrices des Femmes with the help of the UNDP in 1990.

103. Excluding political and labor relations.

104. UNESCO, *Statistical Yearbook 1993* (Paris: UNESCO, 1993).

105. See Victor Bachy, *La Haute-Volta et le cinéma* (Bruxelles: OCIC and L'Harmattan, 1983); FESPACO, *Dixième Fespaco: 18 ans au service du cinéma africain* (Ouagadougou: FESPACO, 1987); FESPACO, *FESPACO: 20ème anniversaire* (Ouagadougou: FESPACO, 1989); and FESPACO, *Festival Panafricain du Cinéma de Ouagadougou* (Paris: Présence Africaine, 1983).

106. Basil Davidson, *The Black Man's Burden: Africa and the Curse of the Nation-State* (New York: Times Books, 1992), 50.

107. Conseil National de la Révolution, *DOP,* 16–17.

# 6

# FOREIGN RELATIONS, OR THE LIMITS OF SOVEREIGNTY

Until 1983 Burkina's foreign relations evolved for the most part unaffected by its successive domestic political crises and changes of leadership. Except in minor details and positions on issues in which Upper Volta bore no influence whatsoever—such as the Arab-Israeli conflict—continuity in foreign policy was the trademark of each regime. The 1960–1983 period formed therefore a diplomatic whole marked by limited pretensions, moderate conservatism, and the ambiguous agency of French and Ivorian interests.

The 1983 revolution brought as much radicalism to foreign as to domestic policy. The CNR attempted to reshuffle Burkina's global and regional alliances and succeeded at least in redefining relations with France. Yet most of the diplomatic changes it introduced were reversed by the succeeding Compaoré regimes of the Popular Front and the fourth republic. The periods from 1983 to 1987 and from 1987 to the present represent thus two additional diplomatic phases, the first one marked by outspoken radicalism and militant foreign policy and the second by a near total return to the policies of the pre-CNR era. The only signs of continuity between the second and third periods have been the maintenance—if not the reinforcement—of a privileged relationship with Libya and the persistence of a certain sense of adventurism, reflected in Compaoré's involvement in the Liberian civil war.

In each period, however, Burkina's diplomatic choices and their consequences have been constrained by the limits history and economics have imposed on its sovereignty. The influence of France, whether direct or mediated through other former French colonies, has been overwhelming, and Burkina's economic vulnerability and its dependence upon foreign aid and opportunities abroad for its migrants have defined the relatively narrow confines of diplomatic realism.

This chapter first reviews the three distinct periods of Burkina's diplomacy. Second, it addresses the specific issues of relations with France and Libya.

## A Small Country with a Limited Role (1960–1983)

Until 1983, foreign policy in upper Volta generally remained on the back burner as the country took "a realistic view of its modest role in world affairs."[1] The main foreign policy concerns in the first two decades of independence were (1) Upper Volta's global positioning with respect to East-West and Arab-Israeli issues; (2) its relationship with Côte d'Ivoire; (3) the issue of border demarcation with neighbors, principally Mali; and (4) the relationship with France. The latter will be dealt with in a section toward the end of this chapter. The first three are briefly reviewed here.

Upper Volta started out under Yaméogo with a generally conservative, Westernaligned foreign policy. It refused to recognize the People's Republic of China and established diplomatic relations with Israel. The first years following Yaméogo's overthrow were little different, partly because Lamizana was uncomfortable in the role of diplomat. Yet in the 1970s Ouagadougou's stance evolved toward slightly more radical and activist positions, especially on issues of African liberation and the Middle East.[2] The drought of 1973 also encouraged Lamizana to diversify his contacts abroad as part of his quest for foreign assistance. In 1973 alone, he attended the OAU summit in Addis Ababa, the Algiers summit of the nonaligned countries, the UN General Assembly in New York, and the francophone summit in Paris and took trips to Tripoli, Rome, and Mecca. The same year, Upper Volta broke off diplomatic ties with Taiwan and established relations with Beijing. Meanwhile, contacts increased with countries of Eastern Europe.

The most remarkable development under Lamizana, however, was the shift toward the Arab world. That Lamizana was Muslim and that he appointed Upper Volta's first Muslim foreign affairs minister were not coincidental to the shift. The first signs of change came in February 1972 when the president visited Egypt, recognized its rights to the Sinai, and declared Upper Volta's support for the UN Security Council Resolution 242, which recognized the Palestinian people's right to self-determination. In September 1973 he severed diplomatic ties with Israel.[3] The pro-Arab shift was completed in 1974 as Upper Volta attended the Islamic summit of Lahore, joined the Islamic Conference Organization (ICO), and declared its recognition of the Palestine Liberation Organization (PLO) as the "sole representative" of the Palestinian people.

Because of their intertwined histories during colonization and as a result of Upper Volta's economic dependence on Côte d'Ivoire, the relationship between both countries has been a paramount dimension of Upper Volta's foreign policy since independence. Yaméogo never questioned his role as an agent of Côte d'Ivoire, and Houphouët-Boigny occasionally did ask Yaméogo to act on his behalf. Such was the case in 1965: As tensions flared between Houphouët and

Guinea's Sekou Touré, the former used Yaméogo to wage a radio war against his Guinean counterpart.[4] It is no wonder that Houphouët was originally displeased with Yaméogo's overthrow and showed little sympathy toward Lamizana, for Lamizana was not part of the "old boys'" network of preindependence RDA politicians, and Houphouët was growing wary of the several military interventions that had occurred in West Africa since Togo had started the trend in 1963.

Yet as irritated as Lamizana and his soldiers were with Houphouët's "avuncular overlordship of Upper Volta,"[5] there was little danger that the country would turn unfriendly toward Côte d'Ivoire if only because it was not fully free in its movements. Ivorian domination of Burkina extended through the Conseil de l'Entente[6] and, from 1970 onward, the Communauté Economique de l'Afrique de l'Ouest (CEAO). In addition, Lamizana was not alone in power. Houphouët's affection for Gérard Kango Ouédraogo, a fellow old-time RDA politician, more than compensated for his dislike of Lamizana from 1971 to 1974. And Ouédraogo reciprocated. In his own words, "some of our RDA friends thought I had cast a magic spell on Houphouët to make him love me like his son. I ... was willing to die for him. I would have thrown myself on him had anyone tried to shoot him."[7] Still, Lamizana had enough power and Ouédraogo was sufficiently absorbed in domestic politics to give Upper Volta a measure of independence from its Ivorian protector: It declined to back Houphouët's policy of dialogue with South Africa. The 1973 severance of ties with Taiwan was also partly designed to display Upper Volta's independence vis-à-vis Côte d'Ivoire.

The third dimension of Upper Volta's foreign relations was the question of where they started: Upper Volta, like other former colonies, was left with vague borders at independence. Although no dispute with Côte d'Ivoire ever surfaced in this respect and although bilateral commissions successfully established frontiers with Ghana and Niger, territorial uncertainties lay at the heart of a substantial conflict with Mali.

Against the background of widespread drought and these uncertain, externally devised borders, Mali and Upper Volta fought a miniwar in 1974 when Mali tried forcibly to claim the 160-km-long Agacher strip, which was known to contain manganese and also believed to harbor oil.[8] Following scattered clashes in December, a war of words was waged on radio waves until Niger's president Seiny Kountché brokered an agreement on 9 February 1975 that provided for the mutual disengagement of military forces from the zone.[9] But the exact demarcation of the border remained in question and would provide further opportunities for conflict.

## A Small Country with Large Ambitions (1983–1987)

The boldness and radicalism of Thomas Sankara's foreign policy contributed much to putting his country on the map in the early 1980s. From a dull and marginal nation at the periphery of the world, Burkina became the new child prodigy of anti-imperialism, Third World pride, and development. For the first time, it broke out of the francophone circle as Sankara traveled the world and caught the attention of the media. Before he died, Sankara was nearing the international status of Ghana's Kwame Nkrumah, Tanzania's Julius Nyerere, and Congo's Patrice Lumumba. His most important speeches had been translated into English and published in the United States.[10] Although his domestic policies contributed to the world's renewed interest in Burkina, it was most of all his self-confident and provocative attitude in foreign affairs that drew attention. Yet upon closer scrutiny, it appears that CNR foreign policy changed quite dramatically around the end of 1985, moving away from bold statements of independence and velleities to export the revolution and toward more pragmatic positions of cordial relations with Western donors and regional conservative powers. This shift in foreign policy paralleled an alteration of the domestic dimensions of the revolution, characterized by Sankara's attempt to open up his increasingly isolated regime.

## Sankara on the World Scene

Only hours after his seizure of power, Sankara proceeded to revolutionize his country's foreign relations. His first step was to accuse France of having sponsored both the CSP-2 regime and his imprisonment at the hands of his army nemesis Somé Yorian on 17 May 1983. President Ouédraogo's dismissal of Prime Minister Sankara and his arrest on 17 May had indeed coincided with the visit to Ouagadougou of President Mitterrand's adviser for African affairs, Guy Penne, who departed the next day. Supporters of the CNR have ever since alleged that Penne came to encourage and supervise this forcible political transition.

The CNR also frequently singled out the United States for its "imperialism," its support of Israel, and its failure to sanction South Africa for its system of apartheid. Because it, too, considered itself a liberation movement, the CNR strongly identified with the PLO, leading to Burkina's concern with the Middle East crisis.[11] To spread his message, Sankara took to every available international platform, not least the General Assembly of the United Nations, where his speech of October 1984 garnered him many Third World supporters.

Although Sankara's rhetoric made him popular among some, it alienated those he blasted. Official Development Assistance (ODA) from France fell from $43.5 million in 1983 to $26.8 million in 1985 and bounced back only when the CNR became tamer and more cooperative in 1986 and 1987.[12] The United States also grew weary of Burkina's rhetoric and its somewhat provocative displays of friendship with Cuba and Libya. Unfortunately for Burkina, the countries whose policies it endorsed (such as Libya, North Korea, Albania, and Yugoslavia) compensated little for the drop in assistance from Western donors. Libya, most notably, repeatedly failed to deliver on its promises of financial assistance.

## Burkina in the West African Region

West African nations, especially those within a same-language group, typically entertain close relations with one another. Age and seniority in power are impor-

*Foreign Relations* 153

tant factors in establishing hierarchies in these relations. Côte d'Ivoire's Houphouët-Boigny, referred to as "the old man" throughout Africa, played a paramount regional diplomatic role from independence until his death in December 1993. As observed earlier, Upper Volta had generally taken on the role as Côte d'Ivoire's satellite.[13] Sankara and the CNR, however, clashed head-on with this tradition and the influence of Houphouët-Boigny and other regional conservative leaders such as Togo's Gnassingbé Eyadéma, Mali's Moussa Traoré, and Niger's Seyni Kountché. But common elements of ideology brought Burkina and Benin closer than before. Most surprising was that revolutionary Burkina established its closest regional ties with Ghana, an English-speaking state. The similar outlook of Sankara and Ghanaian president Jerry Rawlings, their youth, and their comparable accession to power brought the two men together. After the early years of revolution and Sankara's death, though, Rawlings's Ghana evolved in quite a different and more orthodoxically Western direction.

The 1983 coup and its accompanying revolutionary rhetoric caused unease among Ivorian leaders, who were used to a more complacent northern neighbor. Sankara's decision to call on Burkinabè residents abroad to organize into CDRs was received coldly in Abidjan and was countered with official Ivorian hospitality to Burkina's political dissidents and opponents. Sankara broke the tradition of showing respect to Houphouët-Boigny by calling him an agent of French imperialism[14] and refusing to pay the usual courtesy visit to the elder statesman after he seized power. Côte d'Ivoire tried to contain Burkina's influence and succeeded in preventing Sankara's accession to the annual presidency of the CEAO in 1984.

But the overall evolution of Burkina's foreign policy was toward more moderation, and the two countries' antagonism declined over time. Houphouët-Boigny had financial leverage, after all, and Burkina soon understood the impracticality of pursuing an ideologically based foreign policy.[15] In February 1985 the two leaders met for the first time in Côte d'Ivoire, and in March 1986 it was Houphouët-Boigny's turn to pay a visit to Burkina during which Sankara gave him a warm welcome.[16] From then on, relations were normalized between the two countries, and the pattern of Burkina's dependency on Côte d'Ivoire remained essentially unaffected.

Relations turned equally tense with Togo following the 1983 coup but were of lesser significance given Eyadéma's limited regional influence. Then in September 1986 Togo officially accused Burkina and Ghana of involvement in an attempt to overthrow Eyadéma, alleging that Burkina provided Togolese rebels with shelter and a rear base. Bilateral relations remained on edge until Sankara's death in 1987.

While the direct use of force was averted with Togo, Burkina's outspoken desire and threat to export its revolution led to another direct confrontation with Mali in 1985. As observed earlier, a border conflict had existed between Mali and Burkina virtually since the two countries' independence. The zone of contention extended over 160 km between Dionouga in the west and Niger's border in the east (see Map 3.1).

Relations between Traoré's long-standing conservative regime in Mali and the revolutionary leadership in Burkina started out rather well. The CNR had retracted Burkina's veto of Mali's readmission into the UMOA, and both countries had signed an agreement on 16 September 1983 to bring their border dispute before the International Court of Justice in The Hague and to comply with the latter's judgment.[17] Two years later, however, on 25 December 1985, while their case was still pending before the court, the two countries engaged in what is now referred to as the Christmas War. The clashes began because Burkinabè civil servants carrying out the 1985 census went into villages in the contested area, apparently accompanied by CDRs and military troops. Mali accused Burkina of forcing Burkinabè identity cards onto the local residents and harassing local chiefs,[18] and it sent its troops across the border after the alleged Burkinabè culprits. Burkina's defeat was swift. Over five days of fighting, Malian forces penetrated deeply into Burkina's territory, and official casualties eventually amounted to forty-seven. A cease-fire was reached on 30 December through the mediation of Houphouët-Boigny.

Although it was the immediate cause of the conflict, the contested territory actually had little to do with the hostilities. Beyond the census crisis and the border issue, the clashes marked the peak of antagonism between Mali's conservatism and Burkina's radicalism and above all highlighted the limits of Burkina's ambitions. The Christmas War brutally reminded Ouagadougou's leaders that they did not have the means to uphold a self-righteous policy of revolutionary integrity, much less of revolutionary expansion. It emphasized the limitations of Burkina's sovereignty and was a major factor in the CNR's eventual foreign and domestic policy reversal.

Indeed, Burkina had systematically alienated Mali's regimes in the months preceding the skirmish. In July the Malian secretary general of the CEAO, Drissa Keita, had been declared persona non grata in Burkina (where the CEAO is headquartered) for comments critical of the Ouagadougou regime in an interview to the Paris-based *Jeune Afrique* weekly. Burkina's authorities were also investigating corruption involving the CEAO and had spread rumors that Traoré's wife was among those suspected of embezzlement.[19] Most of all, however, Sankara's incendiary speeches had understandably irritated the Malian leadership, which was also seeking diversion from domestic social unrest. At a September 1985 summit of the Conseil de l'Entente in Yammoussoukro in Côte d'Ivoire Sankara had apparently been singled out by his peers as a threat to regional security. On his return from the meeting, the Burkinabè leader publicly threatened Malian authorities with the weapon of his revolution:[20]

> The peoples at our border are also in need of a revolution.... The sister republic of Mali must understand that its happiness will be our happiness.... Burkina's revolution is at the disposal of the people of Mali, who are in need of it, for only the revolution will allow it to fight hunger, thirst, disease, ignorance, and the forces of neo-

colonial and imperialistic domination. Only the revolution will let it free itself. . . . [No] people have a monopoly on the revolution.21

These words were broadcast on radio waves that could be received in Mali and quite conceivably stirred the fears of the Malian leadership. It was only a few more months until the Christmas War exploded. Burkina lost much credibility in its defeat to Mali and was forced, primarily by France and Côte d'Ivoire, to recast the international projection of its revolutionary image in a more moderate tone. Virtually no further mention of exporting the revolution was heard from Burkina afterward.

Niger's leadership did not welcome the 1983 coup either, as it saw its territory progressively circled by radical regimes: Libya in the north, Libyan-occupied Chad in the east, and then Upper Volta in the southwest. Yet relations remained generally cordial, as Burkina never presented a real threat to Niger.

While the CNR ran into trouble with its conservative neighbors, it developed strong ties to its radical counterparts in the region, namely, Benin and Ghana. Burkina entered into a quadripartite agreement among regional progressive regimes—Burkina, Benin, and Ghana—and Libya. Overall, however, Cotonou's orthodox Marxist regime and Mathieu Kérékou's age prevented a sacred union with Benin similar to the one that existed with Ghana. Between Captain Sankara and Flight Lieutenant Jerry Rawlings of Ghana, it was virtually a love affair. Sankara's relative ease of expression in English, fairly uncommon for the head of a francophone state, certainly facilitated the friendly contact. Both men's informality and their taste for reggae and rock music encouraged frequent and occasionally colorful meetings, when they sometimes performed music together. Sankara was also obliged to Rawlings for supporting the 1983 Pô rebellion that led to his release from house arrest and his seizure of power.

Although the two countries ran joint military maneuvers in November 1983 ("Bold Union") and March 1985 ("Team Work"), the personal friendship between their leaders did not materialize into much additional cooperation. Bilateral trade was originally boosted (Burkina's imports from Ghana rose from $1.46 million in 1982 to $5.7 million in 1985) but fell considerably afterward. In addition, Burkina's exports to Ghana remained stagnant throughout the period. Furthermore, even at their peak, in 1985, imports from Ghana still represented only 7.7 percent of those from Côte d'Ivoire, Burkina's traditional regional trading partner.22 The modest successes of the Ghana-Burkina collaboration serve as a reminder of another limit to Burkina's sovereignty: the weight of history, which tends to constrain its foreign relations within the realm of former French colonies. Meaningful and mutually profitable relations are hard to develop outside the francophone region. And as with other dimensions of his revolution, much of what little Sankara had built diplomatically would fall with him upon his death in 1987.

## International Dimensions of the "Rectification" (1987–1994)

### Policy Choices and Diplomacy

With Sankara often more popular abroad than he had been at home, the FP's initial foreign policy concentrated on justifying to the rest of Africa his murder and Compaoré's takeover. The new president visited both conservative and radical capitals in a broad African tour in March and April 1988. Most diplomatic changes occurred, however, with Burkina's neighbors. Relations with Mali quickly and steadily improved, and in May 1988 both countries signed agreements on technical, scientific, and cultural cooperation; livestock migration; transport; and tourism. Shortly after the overthrow of President Moussa Traoré in late March 1991, Burkina's foreign minister, Prosper Vokouma, paid an official visit to Mali's new transitional president, Amadou Toumani Touré. Vokouma expressed his country's solidarity with the Malian authorities and people, presented his condolences to the families of the victims of police violence, and, in an unconventional move, donated a check for CFAFr 50 million ($175,000) to help Malians recover from the abuses of the Traoré regime. Constructive relations continued after the election of Alpha Oumar Konaré as Mali's new civilian president; he paid two official visits to Ouagadougou in 1993, partly dedicated to Burkina's mediation in Mali's Tuareg crisis.[23]

A similar climate prevailed in relations with Togo. Several ministerial-level meetings occurred between the two countries in 1988, and observers remarked on the presence of Eyadéma, the only foreign head of state, at the August 1988 ceremonies for the revolution's anniversary in Ouagadougou, where he heavily praised Compaoré's leadership. At the August summit of the UMOA, Eyadéma supported Compaoré's election as UMOA's new chairman. Finally, Lomé and Ouagadougou were declared sister cities that same August by a friendship pact between their mayors. Some explanation for the intense diplomatic activity between the two countries may be found in Eyadéma's personal desire to obtain the extradition of the Burkinabè suspected of participating in the 1986 coup attempt.

Compaoré was in Togo in December 1988 for a working visit and again in January 1989 to attend the festivities marking Togo's National Day. The dramatic improvement in Burkina-Togo relations continued at an increased pace in 1989. In September Air Burkina opened a branch in Lomé and announced three trips per week between Ouagadougou and the Togolese capital. Compaoré attended celebrations in Lomé again in January 1990. In 1993, after the collapse of Togo's political system, its president, the transitional government, and the opposition signed peace agreements in Ouagadougou under Burkinabè brokerage, and Compaoré announced his decision to commit troops as observers in Togo's elections.

With Ghana, however, relations turned chilly in the immediate aftermath of the October coup but slowly improved afterward. The presence in Ghana of

Compaoré's armed opponent, Boukary Kaboré, poisoned the atmosphere between the two countries until his return to Ouagadougou in April 1991. In January 1990, 120 Ghanaian nationals were deported from Burkina without any official explanation according to Accra's radio, which said that the deported men were mostly traders, artisans, and cobblers. On 26 January the Ghanaian Broadcasting Corporation again accused Burkina of expelling another seventy-four Ghanaian residents. The expulsions were apparently related to Burkina's new policy banning prostitution; many prostitutes in Ouagadougou are indeed Ghanaian. A joint cooperation commission met in May in Ouagadougou. The first since Compaoré's coup, it agreed to institutionalize periodic meetings between the border authorities of the two countries, finalize border demarcations, and revive the bus and air connections between Ouagadougou and Accra.

Compaoré kept a far lower profile in regional arenas than had Sankara. Sankara used to criticize the dysfunction and corruption of many West African organizations. Burkina hosted the June 1989 Economic Community of West African States (ECOWAS) summit, and in the presence of eleven heads of state Compaoré was elected chairman of ECOWAS until June 1990.

On the world scene Compaoré caught observers' attention with an odd outburst when he became the first head of state to visit the People's Republic of China after the events of Tiananmen Square in 1989. Expressing what were widely regarded as outdated opinions, he publicly supported the Chinese authorities, condemning the "reactionary forces" that tried to "force China under the yoke of world capitalism."[24]

Although he had been extremely active establishing his reputation in regional and world forums, it was not until June 1993, after his formal election, that Compaoré went on his first official visit to France. This was perceived as recognition by France of his legitimacy. Having received this French blessing, Burkina's foreign policy grew somewhat bolder and more involved. In November Compaoré continued his role of regional mediator by trying to stage negotiations in Ouagadougou between Niger's government and Tuareg rebels. But the latter failed to show up, and the talks were postponed indefinitely. Burkina had (and still has) a stake in helping solve the Tuareg crisis, for it hosts 30,000 refugees in two camps, one some 45 km from Ouagadougou, the other in Djibo in the north. Discussions between Tuareg rebel movements and the governments of Mali and Niger were held in Ouagadougou under French mediation in September 1994 and led to a truce signed on 9 October between Niger and its rebels.

In December 1993 the parliament approved the sending of forty-three soldiers to Burundi for six months to "protect and monitor activities for the restoration of confidence in that country."[25] The France-Burkina joint commission met in Ouagadougou that same month with Michel Roussin heading the French delegation. Burkina reestablished diplomatic relations with Israel in October 1993 and with Taiwan in February 1994.

## Burkina and the Liberian Civil War

Although in many respects Compaoré's foreign policy has marked a restoration of the pre-CNR diplomacy, stressing links with France and regional responsibilities, the Liberian civil war has provided him with an opportunity for active political and military involvement. Yet as independent-minded and adventurist as it may have appeared, Compaoré's Liberian policy has also partly proceeded from Burkina's agency role vis-à-vis the interests of Côte d'Ivoire, France, and above all Libya.

Liberia's domestic problems took on the dimensions of a civil war around mid-1990 and brought West Africa and the ECOWAS one of their most serious political challenges so far, dragging to the surface the lingering distrust and struggle for regional influence among its members and highlighting Burkina's odd position in regional diplomacy. An August 1990 ECOWAS summit launched the idea of an intervention force to impose a cease-fire among Liberia's warring factions—the government of Samuel Doe, the National Patriotic Front of Liberia (NPFL) of Charles Taylor, and the Independent National Patriotic Front of Liberia (INPFL) of Prince Johnson—and set up an interim government.[26] The force, called the ECOWAS Monitoring Group (ECOMOG), was constituted mostly of Ghanaian and Nigerian troops. Taylor's NPFL opposed the plan, while both the INPFL and besieged President Doe originally backed it up. Taylor had little to gain from the intervention because he already had almost total control of the countryside and believed his conquest of Monrovia and overthrow of Doe was but a matter of time. Doe needed ECOMOG to fend off the advances of the rebels, and the INPFL was militarily weak enough to gain from the intervention.

Among West African nations the main opposition to ECOMOG came from Burkina and Côte d'Ivoire, which made little secret of their sympathy for Taylor (who resided in Ouagadougou before launching his guerrilla campaign). As ECOWAS debated intervention, Compaoré claimed that Liberians wanted to "rid themselves of a system" and that they had to be helped because their "cause [was] just."[27] Niger and Togo were also opposed to ECOMOG but less vocally so. Anglophone states got the upper hand, however, and ECOMOG finally left for Monrovia on 23 August. A few days later, as the first fights took place between the NPFL and ECOMOG, Burkina reaffirmed its opposition by hosting an NPFL delegation in Ouagadougou and restating its opinion that Liberia should be rid of all foreign troops.

Deeds soon followed words. On 29 August 1990 the French daily *Le Monde* reported that loaded with arms and about fifteen Burkinabè soldiers a Libyan airplane had left Ouagadougou and landed at the NPFL-controlled Roberts Field International Airport in Harbel, 56 km east of Monrovia. Rebel sources also told journalists that some of Taylor's partisans were being trained in Libya and Burkina. Later there were further allegations that Burkina had sent some 250 troops to Liberia to fight alongside Taylor's men, that fifty two of them had died in action, and that premiums of up to CFAFr 50,000 were offered to Burkinabè soldiers who volunteered. Foreign journalists also confirmed observing Burkinabè soldiers among Taylor's security personnel.

Compaoré denied all rumors and allegations and pursued an official policy of neutrality and support for attempts at peaceful settlement. He attended the November 1990 ECOWAS summit in Bamako, Mali, that set up an interim government in Liberia with representatives of all factions, called it a "success and a victory for the democratic forces of Liberia," and voiced his hope for the "takeoff of a democratic process . . . despite the foreign interests that have sometimes interfered in the conflict."[28]

Those who believed that the Bamako summit marked a reversal of Burkina's policy were soon to be disappointed. In early April 1991 President Joseph Saidu Momoh of Sierra Leone accused Burkina of cooperating with Liberian rebels from Taylor's NPFL to launch incursions into Sierra Leone to retaliate for Momoh's allowing ECOMOG forces to use the Freetown airport. Momoh declared on national radio that "Burkinabè rebels dressed in blue overalls with red berets operated side by side with the rebels."[29] Once again Compaoré denied the accusations and hardly more than a week later announced that Burkina was considering sending troops to participate in ECOMOG. He did not, however, recuse his support for Taylor. This announcement was followed on 7 May 1991 by the interception by ECOMOG of a German ship flying a Burkinabè flag off the Liberian port of Buchanan, some 150 km east of Monrovia. The ship was transporting rubber from the Firestone plantations under Taylor's control, allegedly to be delivered to Libya in exchange for arms. Compaoré's presence in Tripoli at the time of the interception added substance to the accusations. The arrest of two Burkinabè soldiers in Sierra Leone in late May took all remaining credibility away from Ouagadougou's denials of involvement.

It was thus no surprise when finally, on 9 September 1991, Compaoré admitted having sent 700 troops to Liberia in support of Taylor's NPFL. Yet this was no end to his policy of cultivated ambiguity, for he also claimed that his involvement in the Liberian conflict was now over and that he supported the peace attempts. Sure enough, rumors of additional Burkinabè involvement surfaced again in December, as it was reported that a secret pro-NPFL mercenary army was training in Pô. In March 1992, accusations against Burkina took on a broader dimension, as the Gambia charged the Compaoré regime with trying to destabilize the whole region with Libyan help and training Gambian rebels to invade the Gambia.[30] There followed several months of lower-key Burkinabè diplomacy, including support of ECOWAS resolutions imposing sanctions on Taylor. Nevertheless, in August 1994 the chief of staff of ECOMOG renewed accusations that Burkina was supplying arms to the NPFL, and the Liberian transitional government claimed that some 3,000 Burkinabè-trained mercenaries were fighting alongside the NPFL.[31]

There is little doubt that Burkina has been and remains actively engaged with Charles Taylor in the Liberian conflict and has systematically practiced a diplomacy of deception in this respect since 1990. Few have been fooled, however, and Burkina's Liberian involvement has not been without global consequences. As early as November 1990, the United States blamed Burkina for the continuation of the civil war. The assistant secretary of state for African affairs, Herman Cohen, stressed in a testimony to the Senate subcommittee for African affairs the importance of "singl[ing] out Burkina Faso for sending Libyan arms to one of the factions at war, by air or by road" and for persisting "in contributing to the pursuit of the war even though it had manifestly become a human tragedy."[32] Cohen delivered that message in person to Compaoré in Ouagadougou in December and reiterated U.S. displeasure at Burkina's role in what the Bush administration perceived as the creation of a pro-Libyan coalition in West Africa.

The Libyan dimension of Burkina's intervention provides the key to understanding U.S. reaction. Although it has also actively helped Taylor's rebels, Côte d'Ivoire has not triggered such a response in Washington because the Ivorian policy has been perceived merely as an attempt to counteract Nigeria's regional political influence. If anybody else was behind Côte d'Ivoire's actions in Liberia, it was France, which, like its former colonies, disapproves of a strong Nigeria in West Africa. Burkina's opposition to ECOMOG and its support for Taylor, however, have a Libyan expansionistic dimension that has concerned the United States.

Burkina's problems with the United States are not new. Ever since Sankara took power in 1983, the United States has been wary of Burkina's possible role as a pawn of Libya. Burkina returned the distrust, and in 1986 Sankara put an end to the U.S. Peace Corps program in Burkina. In this perspective Compaoré's problems with the United States are but the continuation of those his predecessor raised through the CNR's friendships with the likes of Libya. U.S. Official Development Assistance, which averaged $33.2 million from 1981 to 1986, fell to $19 million in 1987, $17 million in 1988, $14 million in 1989, and $11 million in 1990.[33] The United States also suspended its military training program agreement with Burkina in December 1991 and eventually recalled its ambassador in November 1992, putting on hold the accreditation of Burkina's ambassador-designate to Washington, Prosper Vokouma. The United States still maintains that Compaoré "actively undermined the ECOWAS peace process by providing military support to Charles Taylor."[34] But the new U.S. ambassador finally arrived in mid-1993. Compaoré has since tried hard to mend fences with the United States.[35]

## Pride and Passion: In the Grip of France

Burkina's relationship with France has been intense and steady but also occasionally difficult and forever tainted by the experience of colonization. Because France created the state of Upper Volta, the ties between the countries lie outside the traditional concept of foreign relations. Still a toddler in a world of independent nations, Upper Volta signed broad-ranging cooperation agreements with France in 1961. The relationship remained in many ways one of tutelage throughout the 1960s and 1970s. To a large extent, it was only Upper Volta's marginality that saved it from further French interventionism. Its recurrent attempts at multiparty politics were also looked upon positively in Paris.

Yet the emotional dimensions that seem to accompany the relations between France and most of its former colonies remained largely absent from its handlings with Burkina in the first two decades of independence. It was only with Sankara's rise to power in 1983 that things took a dramatic turn. From then on, there was enough passion to more than make up for the dullness of the previous years. It should be recalled, first, that France had apparently backed the sidelining of Sankara in May 1983 under the CSP. When he returned to power four months later, Sankara had not forgotten France's apparent involvement in his arrest and was intent on reminding France. The first opportunity he had to make a strong statement of Burkina's new rebellious independence vis-à-vis its former colonizer came in October 1983 at the annual Franco-African summit in Vittel. Sankara showed up in battle uniform with a gun at his waist. Yet France also used the occasion to remind him of his country's marginality. Upon his descent from the plane after landing in France, Sankara was welcomed by Guy Penne, President Mitterrand's adviser for African affairs who had been in Ouagadougou when Sankara was arrested on 17 May and whom the CNR had accused of complicity in this event. Sankara was apparently miffed and refused to attend the inaugural dinner of heads of state. This was the first and last such summit Burkina attended under the CNR. Sankara boycotted the 1984 Bujumbura and following summits, which the CNR referred to as "organizational constraints inherited from the colonial period."[36] Beyond symbolism, the CNR actually demanded and obtained a renegotiation of the essential cooperation agreements linking France and Burkina; a new cooperation convention was signed in February 1986.

As was the case with Côte d'Ivoire and Mali, the period covering the end of 1985 and the beginning of 1986 also marked a change in Burkina's policy toward France. Revolutionary rhetoric and denunciations of French imperialism and neocolonialism were toned down, and Sankara traveled to Paris for an environmental conference in February 1986, taking that opportunity to meet with President Mitterrand at the Elysée palace. In addition, while Sankara still declined to show up for the 1986 Lomé Franco-African summit, he invited the French president to stop by Ouagadougou for an official visit on his way back to Paris, which Mitterrand did.[37] That such changes occurred after French development assistance had fallen by about 50 percent between 1983 and 1985 is probably no coincidence. And indeed, Burkina's leadership seems to have been rewarded for its policy shift. French ODA climbed back from $26.8 million in 1985 to $40.8 million in 1986 and $66.8 million in 1987, the CNR's last year.[38] Part of the gains, however, were due to the appreciation of the French franc over the same period.

Given the improved relations with Burkina and Mitterrand's apparent personal fondness for Sankara, France's first reaction to the Compaoré coup was cool, although it marked more a disapproval of the methods used than of political change per se. In December, however, Burkina resumed its participation in Franco-African summits, attending the latest in Casablanca. Despite France's displeasure with Compaoré's continued friendship with Libya, Paris eventually resumed its military cooperation and helped structure Compaoré's personal security.

But it was not until the Ouagadougou visit of French cooperation minister Jacques Pelletier in January 1990 that Franco-Burkinabè relations were fully normalized (Burkina still accused France of involvement in the Lingani-Zongo alleged coup attempt). Pelletier presided over the meeting of the mixed cooperation commission that had been postponed four times since 1987, and France wrote off a substantial chunk of Burkina's debt while making fresh financial commitments. In June Compaoré also made the trip to the Franco-African summit in La Baule, France, where Mitterrand pledged support to regimes embarking on democratic reforms. He later granted Burkina CFAFr 250 million toward the cost of the 1991 constitutional referendum. Overall, French ODA soared to new heights under the Compaoré presidency. From $54.8 million in 1988, it climbed to $80.2 milion, $93.8 million, $125.9 million, and $133.4 million in the four following years.[39]

In September 1994, in a move to further ingratiate itself with Paris, Burkina agreed to host twenty Algerian Islamist deportees from France. This, together with Compaoré's mediation successes in Togo and in the Tuareg crisis, led to increased status for the president at the November Franco-African summit in France and to a reward in the form of the announcement that the 1996 summit would be held in Ouagadougou.

## The Enduring Libyan Connection

The peculiar relationship between Burkina and Libya began under the CSP regime of Jean-Baptiste Ouédraogo at the initiative of Sankara and hardly changed under the CNR or FP. Barely more than two months after having been appointed prime minister of the CSP, Sankara visited Libya in March 1983 and received Muammar Qadhafi in Upper Volta in April. While in Ouagadougou, the Libyan leader encouraged Upper Volta—which he called his second country—to continue the expansion of the "anti-imperialist front in Africa."[40] There is little controversy that Sankara's Libyan policy, and specifically Qadhafi's visit, contributed to his ousting from the CSP a month later. But if Libya had cost Sankara his portfolio, it also tried to bring him back to power thereafter. While he was under house arrest, it is reported that the Libyan regime set up the Voice of the Voltaic Revolution radio station in occupied northern Chad and broadcast to Ouagadougou calls for the overthrow of the CSP.

Sankara duly renewed the relationship when he returned to power in August. However, geopolitical reality caught up with both leaders' expectations in the course of the CNR presidency, and the Burkina-Libya alliance ended up delivering little either in terms of support for Burkina or control for Libya. The CNR support for the Polisario Front in Western Sahara and for the independence of the Arab Saharaoui Democratic Republic, as well as its opposition to foreign intervention in Chad—whether French or Libyan—helped cool off the Libyan connection. So did a progressive intensification of relations between Burkina and Algeria, which the CNR perceived as more generous and less of a diplomatic liability. Nevertheless, Burkina's relations with Libya retained a special character, and the two leaders exchanged several visits between 1983 and 1987, including a trip to Ouagadougou by Qadhafi in December 1985 during which he unsuccessfully attempted to proclaim Burkina a *jamahirya*, or Islamic republic.[41]

While most observers expected the Libyan connection to end with the death of Sankara, the opposite took place: Compaoré maintained and reinforced the friendship. He visited Libya in January 1988, only three months after assuming power, and in April sponsored the creation of a joint bank (the Banque Arabe-Libyenne du Burkina) headquartered in Ouagadougou. Officers from both countries' armies also launched a Burkinabè-Libyan Friendship and Solidarity Society. Barely four months after his January visit, Compaoré stopped again in Libya in May, on his way to the Addis Ababa OAU summit and spent forty-eight hours in Tripoli on his way back a few days later. He had a long meeting with Qadhafi during his second stay. Earlier, in April, Burkina Faso had expressed its support for Libya on the occasion of the second anniversary of the 1986 U.S. air raid on Bengazi. In its statement the FP mentioned that the Burkinabè revolution was still "an integrated part of the world anti-imperialist movement" and remained in solidarity with "the peoples' liberation struggles against imperialism, apartheid, and Zionism."[42] It was likely that the FP was thereby trying to get Colonel Qadhafi to fulfill some of the many promises of financial assistance he had made in the previous few years.

Washington recalled its ambassador to Burkina in January 1989 for consultations after Compaoré described the U.S. downing of two Libyan planes on January 4 as "U.S. imperialist aggression" and a "terrorist act" and expressed support for Qadhafi. Compaoré's broadcast message was strongly worded, especially in view of the United States' position as Burkina's second largest aid supplier after France. The U.S. ambassador returned to Ouagadougou in early February. The U.S. action may have impressed Burkina and further encouraged an already ongoing evolution; at any rate, this was one of the last occasions of CNR revolutionary rhetoric.

But the substance of the relationship with Libya remained unaltered. Burkina and Libya had signed a cooperation agreement in November 1988, only a few weeks before the planes' downing. It called for consolidation of economic and technical cooperation; increased cooperation in air transport, information, culture, and commercial exchange; the construction of factories; and the establishment of a joint company to set up gas stations.

Compaoré again traveled to Tripoli in August 1989 for ceremonies making the twentieth anniversary of the Libyan revolution. On his way back from the La Baule Franco-African summit in June 1990, he headed for Tripoli for an eight day private visit. In August 1992, after driving overland from Tunisia because of the ban on flights to Libya, Compaoré held two meetings with Qadhafi. Libya announced unspecified debt-relief measures and promised financing for the construction of a 100-bed maternity hospital in Ouagadougou.[43]

Since then the evidence of joint actions by Burkina and Libya in the Liberian civil war have further confirmed the maintenance of a special relationship between the two countries and of an adventurist streak in Compaoré's apparent conservatism.

## Notes

1. *Africa Contemporary Record (ACR)*, vol. 4 (New York: Africana Publishing, 1971–1972), B703.

2. Ibid., B703–704.

3. *ACR*, vol. 6 (New York: Africana Publishing, 1973–1974), B788–789.

4. See *Jeune Afrique Plus*, 8 (June 1984): 63.

5. *ACR*, vol. 6, B789.

6. See Virginia Thompson, *West Africa's Council of the Entente* (Ithaca, N.Y.: Cornell University Press, 1972).

7. *ACR*, vol. 4, B704.

8. For details, see *Le Monde*, 18–20, 22, 23, and 26 December 1974.

9. *ACR*, vol. 7 (New York: Africana Publishing, 1974–1975), B710–713.

10. See Thomas Sankara, *Thomas Sankara Speaks: The Burkina Faso Revolution, 1983–1987*, trans. Samantha Anderson (New York: Pathfinder, 1988).

11. This is also at the root of its support for the Polisario Front in Western Sahara. Sankara was the first head of state to go on an official visit to the Saharaoui Republic in April 1984.

12. OECD, *Geographical Distribution of Financial Flows to Developing Countries: Disbursements, Commitments, Economic Indicators, 1985/1988* (Paris: OECD, 1990).

13. The relation between Côte d'Ivoire and Upper Volta was qualified as "subimperialism" by French Africanist Pierre-François Gonidec, *Les Systèmes politiques africains* (Paris: Librairie Générale de Droit et de Jurisprudence, 1978), 330.

14. *Jeune Afrique*, 1199–1200 (28 December 1983–4 January 1984): 77.

15. See Y. A. Fauré, "Ouaga et Abidjan: Divorce à l'africaine? Les raisons contre la raison," *Politique Africaine*, 20 (December 1985): 78–86.

16. See *Carrefour Africain*, 928–929 (4 April 1986): 20, for the text of Sankara's speech in honor of Houphouët-Boigny in which he lauds the latter for his role in the period leading to independence.

17. See *Carrefour Africain*, 793–794 (2 September 1983): 16, and 797 (23 September 1983): 17.

18. *Marchés Tropicaux*, 2095 (3 January 1986): 3.

19. See *Africa Confidential*, 29 January 1986, 5.

20. Mali is not a member of the Conseil de l'Entente but had probably communicated its concerns to members sympathetic to it.

21. *Carrefour Africain*, 901 (20 September 1985): 14.

*Foreign Relations* 165

22. International Monetary Fund, *Direction of Trade Statistics Yearbook*. This does not take into account, however, the occasional use of barter between both countries.

23. Tuareg communities of northern Mali and Niger revolted against their states in the 1990s, resorting to armed insurgency and terrorist actions to support claims for autonomy.

24. Economist Intelligence Unit (EIU), *Country Report: Togo, Niger, Benin, Burkina*, 4, 1989.

25. Ibid., 1, 1994.

26. For a detailed description of Liberia's civil war, see Christopher Clapham, "Liberia: Recent History," in *Africa South of the Sahara 1995* (London: Europa Publications, 1994), 528–532.

27. EIU, *Country Report*, 3, 1990.

28. Ibid., 1, 1991.

29. Ibid.

30. Ibid., 2, 1992, 42.

31. Ibid., 4, 1994, 47.

32. Ibid., 1, 1991.

33. OECD, *Geographical Distribution*.

34. EIU, *Country Report*, 1, 1993, 46.

35. See "Mission to Ouagadougou," *World View*, 6, 3 (Summer 1993): 22–23.

36. CNR declaration of December 1984, reprinted in Pierre Englebert, *La Révolution burkinabè* (Paris: L'Harmattan, 1986), 259.

37. For excerpts of both president's speeches on this occasion, see *Le Monde*, 19 and 20 November 1986.

38. OECD, *Geographical Distribution*, 72–73.

39. Ibid.

40. *Carrefour Africain*, 777 (6 May 1983): 7.

41. See ibid., 913 (13 December 1985): 12.

42. Agence France Presse, *Bulletin Quotidien d'Afrique*, April 1988.

43. EIU, *Country Report*, 4, 1992, 44.

# 7

# CONCLUSION
## *A Historical Parenthesis or the Foundations of Enduring Statehood?*

The future of Africa's nation-states is uncertain. Throughout the continent the modern state is stretched to or beyond its limits. In some cases—as in Angola, Liberia, Rwanda, Sudan, and Somalia—it simply provides a context for protracted and amazingly cruel civil wars. In others, like Mauritania, it supplies the stage and the instruments for the ruthless domination of one or more groups by another. In many more, not least Zaire and Nigeria, the state (or what is left of it in the case of Zaire) is a predatory enterprise of patrimonial exploitation and appropriation of the assets of the many for the benefit of a few. Most anywhere else south of the Sahara, it often seems to be a parody of the model it was founded upon, kept alive by the artifice of international sovereignty and transfusions of foreign aid.

Burkina is among but a few African countries to have escaped the worst dimensions of this predicament. To be sure, its sovereignty is mostly juridical, and it is highly dependent on foreign economic assistance. Yet it has now experienced thirty-five years of civil peace during which the state has treated its citizens with only mild inequality and has provided few with the means of personal economic advancement. In its nature, however, Burkina is no different from other postcolonial states in Africa. As this book has shown, it is the happy combination of several elements that has made Burkina a relative African success story. Should circumstances change, it remains as vulnerable as other African countries.

## A Relative Success Story

Why has Burkina performed relatively well by African standards with respect to civil peace, governance, and economic growth? First, it has benefited from the absence of ethnicity—defined as the political expression of ethnic identity—among its people. A multiethnic situation such as Burkina's is indeed not in and of itself always a factor of instability and underdevelopment, as illustrated, for example, by the history of the United States. For multiethnic circumstances to become a liability to development and stability, ethnicity must compete with the state for allegiance and undermine the state's legitimation, as has been the case with Yugoslavia in the first half of the 1990s. This element of alternative legitimacy, fairly widespread in Africa, has not occurred in Burkina for several reasons. First, the French colonizer effectively crushed the political structures of the main ethnic group, the Mossi, and deeply compromised its leadership by forcing its collaboration with the colonial enterprise (this did not happen with the same intensity in every French colony). Second, many other ethnic groups, principally in the western part of the country, never organized as political communities and, as such, never generated loyalties alternative to the state. Third, the aversion of Mossi chiefs to formal colonial education kept their children away from school and allowed for the emergence of a "modernized" political elite made of commoners without allegiance to ethnic authorities. Finally, the absence of interethnic conflict has created a climate of social peace and relative tolerance and has prevented the emergence of ethnic competition for the levers of power. As a result of these four factors, the historical legitimacy of the state in Burkina has not been deeply challenged, and the situation has not degenerated into social conflict or civil war; rather, the consequences have been confined to comparably innocuous political instability.

The lack of significant challenge to the legitimacy of the state also accounts in part for Burkina's relatively good governance, characterized by the absence of patrimonialism (i.e., leaders' use of the state's economic resources for personal ends) and of corruption and by the presence of a surprisingly efficient civil service. Because Burkina's political leaders have few other loyalties, they have tended to run the state with probity rather than as a source of personal wealth. In addition, the country's limited resources have given them few opportunities for large-scale deviance. As for civil servants, in addition to having inherited a culture that stresses the value of hard work—as a matter of mere survival—they, too, have had few institutional incentives to use the state as a redistribution mechanism, despite the prospects for rent seeking created by the state's involvement in the economy. This has come as the result of the lack of both strong ethnic identification and interethnic conflict, which has prevented the development of an attitude of "them" versus "us" conducive to abuse of privileges.

Burkina's record of economic growth owes much to circumstances outside government. Although nature endowed Burkina with little, the specter of drought, at

*Unsteady statehood: Human pyramid on a motorbike, part of the military parade for the anniversary of the revolution (1985 or 1986). Courtesy of Sidwaya.*

least, has not haunted the country since the early 1980s. And the inhabitants of this young country have great resilience. Their capacity to diversify their economic activities in the face of agricultural adversity has allowed Burkina to weather recessions and droughts with less human suffering than in many other places. Migration, which is nothing other than the geographical diversification of income risks, has been the essential component of individual economic strategies and has contributed much to Burkina's performance.

Nevertheless, whereas many other African countries have squandered their meager assets through insensible policies, the relative soundness of many of Burkina's economic choices has helped it make the best of its limited resources. Since the early 1980s Burkina has successfully managed a policy of economic diversification to move it away from its dependence on cotton for foreign exchange revenue. Gold became the second export in the mid-1980s, and the recent beginning of manganese mining is moderately promising. Development projects have also been generally well conceived, and the government has attempted to centralize and give a sense of direction to the many grass roots projects undertaken by NGOs around the country. Finally, the government has so far been one of the very few in the Franc Zone to have successfully dealt with the consequences of the 100 percent devaluation of January 1994. Inflation has been contained while producer prices have been allowed to rise, reflecting the increased competitiveness of agricultural exports. In addition, livestock exports are now believed to have regained their regional market share previously lost to Latin American and European meat.

## Lingering Political, Ethnic, and Economic Threats

Yet for all its social, governmental, and economic successes, Burkina remains susceptible to most of the plagues that recurrently afflict the continent. First, its polity remains highly unstable, and the sources of its instability are several. Unlike many other former colonies in Africa, Upper Volta reached independence with none of its early political leaders in office: Coulibaly and Zinda Kaboré were dead, Nazi Boni was marginalized, and Houphouët-Boigny was running Côte d'Ivoire. As a result, Upper Volta did not benefit from the unifying and stabilizing presence of a sort of "father figure" of independence. President Yaméogo was an obscure politician who first came to politics in 1957, just three years before independence, and reached the pinnacle of power without popular mandate and only after substantial parliamentary maneuvering. The political elite's lack of historical and popular grounding was no doubt a factor in the emerging pattern of instability. Yaméogo was wiped out of office in a matter of hours, with virtually no resistance and in a dramatic display of the superficiality of his power, best illustrated by his lack of control over instruments of statehood such as the military. The ease with which regimes are overthrown has since remained a characteristic of Burkina's political life.

Second, the colonial authorities' repression of the Voltaic RDA after World War II (much stronger than in other West African colonies), together with its encouragement of a distinct Mossi political movement, contributed to preventing the emergence of a strong nationwide political party in Upper Volta. Elsewhere in former French West Africa, branches of the RDA constituted the cement of national political unification and provided a measure of stability, although often at the price of democracy. Ironically, therefore, the repression of radical politics by the Voltaic colonial administration, while spreading the seeds of instability, also saved Burkina from the throes of single-party politics and gave its polity a distinct democratic dimension.

Third, the repression of political parties allowed the trade unions to flourish. Outside the political system but staffed by insiders, unions have been used for political ends rather than labor issues and have contributed more than any other group to regime instability. Finally, Burkina's economic volatility has stimulated its political inconstancy by repeatedly forcing governments to curtail the privileges of the urban classes on which the state's power relied.

Instability has manifested itself in several ways. For one, politicians have made a joke of parliamentary regimes, resorting to alliances and majority reversals for no other reason than to affect the distribution of ministerial portfolios to their advantage. This pattern, in which individuals matter more than parties, programs, and ideologies, started in 1957 when Joseph Conombo and his supporters, displeased at their lack of importance in the governmental coalition, quit the Coulibaly government to create a new party and a new majority with former opposition parties and tried to force Coulibaly to resign. The only way Coulibaly avoided this fate was by recruiting four opposition parliamentarians into his own party, thereby reobtaining a majority. A similar story of parliamentary deadlock between Gérard Kango and Joseph Ouédraogo in 1974 led to the return of the military. Since the 1990s the participation of Herman Yaméogo, the son of the late president, in several Compaoré governments has proceeded from the same logic of personal political advancement. With a few exceptions, therefore, Burkina's political parties remain as fluid and immaterial as ever.

Instability has also taken the form of military factionalism. Until recently, most political currents were represented in the military, and the latter loyally reproduced the divisions of the political society: The accession to power of Saye Zerbo was indirectly that of Joseph ki-Zerbo; Somé Yorian's faction possibly wished to reinstate Maurice Yaméogo; and with the CNR it was the turn of the radical leftist parties to gain access to the state. Today it is unclear, however, whether this dimension of political instability persists, for both the CNR and the FP have thoroughly cleansed the military of its political pluralism. The CNR provided for the killing or eviction of most conservative or pro-ki-Zerbo officers, such as Somé Yorian and Saye Zerbo. Compaoré continued the purge, this time with respect to other leftist factions (the 1987 coup saw the death of Sankara and his close aides, and the 1989 murders of Lingani and Zongo completed the elimination of the "old wave" of military revolutionaries). It is therefore possible that the military has ceased to harbor factions and ideological divisions and that its propensity to stage coups has been reduced. There remains a substantial unknown, however, in the person of Diendéré, who has adopted a low profile since he was rumored to have gained substantial political ascendancy around 1989.

A third indication of political unsteadiness is provided by the systematic fuzziness of the political orientation of Burkina's regimes and the secrecy of politics. To this day, for example, it is no easy task to assess Compaoré's regime and the relationship between the president's now avowed liberalism (after years of claiming socialist integrity) and the radicalism of the majority party, the ODP/MT. The ODP/MT is certainly not what is was a few years ago, when under the chairmanship of Clément Oumarou Ouédraogo it professed to unite Burkina's remaining communist groupings. Nevertheless, although it dropped all references to Marxism-Leninism in the early 1990s, it still counts pure-blood revolutionaries among its members. The actual political objectives of these individuals, as well as the balance of power between them and the president, remain mostly unknown and may contain the seeds of future instability.

The democratic claim of Burkina's current regime also begets examination. Despite its appearances and historical precedents, Compaoré's Burkina is less a democracy than a dictatorship, however mild, under the guise of democracy. Compaoré's manipulation of politics, the hegemonic control of parliament by the ODP/MT, the military foundations of the current regime (both Compaoré and Yé came to politics through coups and have only formally severed their military ties), the repression of student politics, and the signs that torture remains in use all indicate that Burkina may only be paying lip service to democratization.

President Compaoré is a consummate politician, aware of the desire of donor countries to see more signs of democracy in Africa. He has displayed savvy and overcome considerable odds in strengthening his regime and liberalizing its economy. Yet he has much blood on his hands, including that of his friends and former comrades-in-arms, and it seems he is deterred by little in his quest for power. He crushed opposition attempts to organize a national conference before the 1991 presidential elections. To be fair, he is helped by an unorganized and inarticulate opposition where cleavages dominate over consensus. Although it remains conceivable that Compaoré may in his turn be faced with the state's incapacity to bend civil society and end up the victim of a swing in regimes, he has given signs of a surprising propensity for political longevity. In addition, in a region now deprived of Houphouët-Boigny's influence and with Eyadéma's government in shambles in Togo, the active French endorsement of Compaoré's regime (won through Burkina's civil peace and its helpfulness to France's African policy) has helped raise Compaoré's political life expectancy and shielded him from pressures for further democratization.

A further potential danger, ethnicity, still threatens Burkina. Should the social and cultural picture change—and the heavy Mossi migrations westward due to population pressure may well spur a redefinition of ethnic identities and relations—the ingredients for an ethnic explosion would be present. What has until now been a case of instability in the polity could turn into a serious conflict in the society. The good rains since the mid-1980s have allowed the Mossi to carry out their agricultural colonization of the west with little resistance. But a recurrence of drought or exhaustion of the available land would make resources scarcer and could well present a challenge to Mossi dominance.

Finally, the recent clement weather has also temporarily hidden the profound contradiction between the apparent absence of profit motive and accumulation among many Burkinabè and their state's modern economic rationality and desire for growth. Because their environment and culture have stressed redistribution over accumulation, few of Burkina's farmers have grown wealthy since independence. Meanwhile, their state has based its own growth on the assumption that it could appropriate an agricultural surplus. As a result, no fewer than 30 percent of Burkina's migrants cite taxes as their reason for leaving; in a manner reminiscent of colonial patterns of exploitation, the state is draining its people of their sparse income. It is dubious whether any long-term sustainable development can arise within this context.

## Toward Steady Statehood and Sustainable Development?

How can the lasting constraints of instability, authoritarianism, ethnicity, and poverty be alleviated? No doubt this question can be addressed on many levels, and democratization and economic liberalization are part of any answers. But one essential condition for long-term steady and legitimate statehood in Burkina, as elsewhere in Africa, lies in the state's capacity to listen to civil society and to accommodate rather than intimidate it. Most observers agree that a state-society dichotomy exists in Africa. Yet most seem to believe the state should mold its societies into a nation. In my opinion, this is only a recipe for short-term repression, not long-term development. The real solution lies in the endogenization of the state by society, for first things must come first. Remember that the French certainly never considered Upper Volta to be a meaningful national entity. They created it in 1919 mainly for political reasons: to increase their control of the rebellious western peoples and to use the complacent Mossi as a taming force. Once that was taken care of, they had no qualms taking it apart for economic reasons. And when it was reconstructed in 1947, it was again for political reasons, this time to put a brake on the spread of the RDA in West Africa. Never were the interests of a hypothetical Voltaic nation identified or taken into account. It is hard to take such a state seriously. In fact, the African postcolonial state may be able to foster development only if it becomes the legitimate product of domestic social, political, and economic history.

The practical question is whether African states in general, and Burkina for our purpose, can generate their own legitimacy or whether they will have to give way to other institutions. In some countries, such as Liberia in West Africa, Zaire in Central Africa, and Somalia in East Africa, the postcolonial state (if Liberia can be so qualified) has finally begun to crumble under the repeated military, political, ethnic, and economic assaults of societies (it is interesting that at different times foreign states have intervened to try to prevent the deliquescence of the Liberian, Somalian, and Zairean states). In countries such as Burkina, however, it may not be necessary for violence and human suffering to reach such levels before the state-society adjustment occurs. It is possible that in thirty-five years of independence, the Burkinabè state has had enough time and has been sufficiently benevolent toward its people to become socially endogenous and to generate sufficient legitimacy to successfully withstand social challenges. It is conceivable that the nationalistic ideology fostered by the 1983 revolution and the strong identity that has come out of it for Burkina, in comparison to Upper Volta, have finally laid the foundations for a nation. The state may thereby have created its own historicity and national identity, helped by the weakness of ethnic structures, the young people of Burkina, and increased urbanization. Witness the continued annual celebration in Ouagadougou of the anniversary of the August 1983 revolution, despite the radical change of regime.

If Burkina has indeed established its own identity, then the period since 1960 will have been of the utmost historical significance and will have witnessed the birth of a nation-state. If not, Upper Volta and Burkina will be in parentheses in the long-term history of the region. In this latter case, as in other African countries where independence has failed to bring about legitimacy, it is hard to imagine what alternative forms of statehood may eventually be generated. It would border on naive romanticism to believe that ethnic institutions could again take up the power and authority they had about a century ago. Yet incapable as they may be of bearing the whole burden of contemporary legitimacy, customary institutions still would be a part of the solution. One could conceive of a system where ethnic institutions would take over state power in the areas where they have remained potentially functional, such as land management, administration of justice, and organization of local product, land, and labor markets. This recommendation thus involves a decentralization of state functions at the level of legitimate customary institutions in a sort of loose federalism where the peculiarities of different cultures and social organizations would be respected. In other fields, such as the provision of credit, health care, primary education, and agricultural extension services, where neither the state nor ethnic authorities hold a comparative advantage and where there may be a legitimacy vacuum, there is room to set up modern, community-based institutions. The widespread success of NGOs throughout Africa bears testimony to the demand for their services and to their capacity to generate legitimacy.

As community-based associations were fostered, the functions of the state could be reduced and the location of its authority displaced. The state could concentrate on its traditional roles of macroeconomic management, provision of public goods (infrastructure, higher education, etc.), and alleviation of market failures. In order to prevent excessive state growth and to benefit from potential economies of scale, there could be supranational state integration with repeal of current borders within the new state. The UMOA or other regional organizations could provide a basis for such integration. The resulting system of loose federalism with strong community public management could be equated to a constitutional share-contract where a remote central government and local communities share both a stake in political legitimacy and the dividends of economic outcomes.

Such arrangements would require nothing short of a revolution and the dedication of disinterested leaders. But African and Burkinabè history is not devoid of inspired leaders and attempts at supranational integration. Those like Senghor and Modibo Keita, who lobbied for African federalism and set up the Federation of Mali in 1960, had a remarkably early perception of the legitimacy deficit of the states they were about to inherit. That they failed does not invalidate their project's raison d'être. It only serves as a reminder of the tremendous obstacles that lie ahead.

# *Bibliography*

Adloff, Richard. *West Africa: The French-Speaking Nations Yesterday and Today.* New York: Holt, Rinehart and Winston, 1964.

Ammi-Oz, Moshe. "L'Installation des militaires voltaïques (I)." *Revue Française d'Etudes Politiques Africaines—Le Mois en Afrique,* 152–153 (August-September 1978): 59–79.

———. "L'Installation des militaires voltaïques; (II)." *Revue Française d'Etudes Politiques Africaines—Le Mois en Afrique,* 154 (October 1978): 85–97.

Ancey, Gérard. *Monnaie et structures d'exploitations en pays mossi, Haute-Volta.* Paris: ORSTOM, 1983.

Andriamirado, Sennen. *Il s'appelait Sankara: Chronique d'une mort violente.* Paris: Jeune Afrique Livres, 1989.

———. *Sankara le rebelle.* Paris: Jeune Afrique Livres, 1987.

Assises Nationales sur le Bilan de Quatre Années de Révolution. *Documents finaux.* Ouagadougou: Front Populaire, 1988.

Audouin, Jean. "L'Evangélisation des Mossi par les Pères Blancs: Approche socio-historique." Doctoral dissertation, Ecole des Hautes Etudes en Sciences Sociales, Paris, 1982.

Audouin, Jean, and Raymond Deniel. *L'Islam en Haute-Volta à l'époque coloniale.* Paris: L'Harmattan and INADES, 1978.

Augustin, Jean-Pierre, and Yaya K. Drabo. "Au sport, citoyens!" *Politique Africaine,* 33, (1989): 59–65.

Ayittey, George B. N. "Indigenous African Systems: An Assessment." In *Background Papers: The Long-Term Perspective Study of Sub-Saharan Africa,* vol. 3: *Institutional and Sociopolitical Issues.* Washington, D.C.: World Bank, 1989, 22–31.

Bachy, Victor. *La Haute-Volta et le cinéma.* Bruxelles: OCIC and L'Harmattan, 1983.

Badie, Bertrand. *L'Etat importé: L'occidentalisation de l'ordre politique.* Paris: Fayard, 1992.

Balesi, Charles John. *From Adversaries to Comrades-in-Arms: West Africans and the French Military, 1885–1918.* Waltham, Mass.: Crossroads Press and African Studies Association, 1979.

Balima, Albert S. *Genèse de la Haute-Volta.* Ouagadougou: Presses Africaines, 1970.

———. "L'Organisation de l'empire Mossi." *Penant-Revue de Droit des Pays d'Afrique,* 74, 703–704 (October–December 1964): 477–498.

Bamouni, Babou Paulin. *Burkina Faso: Processus de la Révolution.* Paris: L'Harmattan, 1986.

———. *Principes d'action révolutionnaire.* Ouagadougou: Direction Générale de la Presse Ecrite, 1984.

Banque Centrale des Etats de l'Afrique de l'Ouest. "Burkina: Statistiques économiques et monétaires." *Notes d'Information et Statistiques.*

Barral, Henri. *Les Populations nomades de l'Oudalan et leur espace pastoral.* Paris: ORSTOM, 1977.

Bassolet, François D. *Evolution de la Haute-Volta*. Ouagadougou: Imprimerie Nationale, 1968.
Bates, Robert H. *Markets and States in Tropical Africa: The Political Basis of Agricultural Policies*. Berkeley: University of California Press, 1981.
Baxter, Joan, and Keith Sommerville. "Burkina Faso." In Chris Allen et. al., *Benin, the Congo, Burkina Faso: Economics, Politics, and Society*. London: Pinter Publishers, 1989, 237–300.
Bayart, Jean-François. "Civil Society in Africa." In Patrick Chabal, ed., *Political Domination in Africa: Reflections on the Limits of Power*. Cambridge: Cambridge University Press, 1986, 109–125.
_____. *La Politique africaine de François Mitterrand*. Paris: Karthala, 1985.
Bayart, Jean-François, Achille Mbembe, and Comi Toulabor. *Le Politique par le bas en Afrique noire: Contributions à une problématique de la démocratie*. Paris: Karthala, 1992.
Bazié, Jean-Hubert. *Chroniques du Burkina*. Ouagadougou: Imprimerie de la Presse Ecrite, 1986.
Benabdessadok, Christine. "Femmes et révolution, ou comment libérer la moitié de la societé." *Politique Africaine*, 20 (December 1985): 54–64.
Benoît, Michel. *Introduction à la géographie des aires pastorales soudaniennes de Haute-Volta*. Paris: ORSTOM, 1977.
_____. *Nature peul du Yatenga: Remarques sur le pastoralisme en pays mossi*. Paris: ORSTOM, 1982.
_____. *Oiseaux de mil: Les Mossi de Bwamu (Haute-Volta)*. Paris: ORSTOM, 1982.
Berg, Elliot. "Mobilization of the Private Sector in Burkina Faso: Report Prepared for the Ministry of Plan and the Chamber of Commerce, Agriculture and Industry." Ouagadougou, December 1985.
Bhatia, R. J. *The West African Monetary Union: An Analytical Review*. Occasional Paper No. 35. Washington, D.C.: International Monetary Fund, 1987.
Binger, Louis. *Du Niger au Golfe de Guinée par le pays Kong et le Mossi, 1887–1889*. Paris, 1892.
Binswanger, Hans, and John McIntire. "Behavioral and Material Determinants of Production Relations in Land-Abundant Tropical Agriculture," *Economic Development and Cultural Change*, 36, 1 (October 1987): 73–99.
Binswanger, Hans, John McIntire, and Chris Udry. "Production Relations in Semi-arid African Agriculature." In Pranad Bardhan, ed., *The Economic Theory of Agrarian Institutions*. Oxford: Clarendon Press, 1989, 122–144.
Bonaventure, Traoré. "Canette de bière ou calebasse de dolo." *Le Monde Diplomatique*, 360 (March 1984): 10–11.
Boni, Nazi. *Le Crépuscule des temps anciens*. Paris: Présence Africaine, 1962.
Bonnet, Doris. *Le Proverbe chez les Mossi du Yatenga, Haute-Volta*. Paris: Société d'Etudes Linguistiques et Anthropologiques de France, Agence de Coopération Culturelle et Technique, 1982.
Borders, William. "In Upper Volta, No One Fears the Tribal Sun King Anymore." *New York Times*, 17 August 1971, 14.
Borella, François. *L'Evolution politique et juridique de l'Union Française depuis 1946*. Paris: Librairie Générale de Droit et de Jurisprudence, 1958.
Bosch, Ellie. *Les Femmes du marché de Bobo: La vie et le travail des commerçantes dans la ville de Bobo-Dioulasso au Burkina Faso*. Leiden: Rijksuniversiteit te Leiden, 1985.

Boughton, James M. "The CFA Franc: Zone of Fragile Stability in Africa." *Finance and Development*, 29, 4 (December 1992): 34–36.
Bourdier, Jean-Paul, and Trinh T. Minh-ha. *African Spaces: Designs for Living in Upper Volta*. New York: Africana Publishing, 1985.
Brasseur, Gérard, and Guy Le Moal. *Cartes ethno-démographiques de l'Afrique Occidentale Française*. Dakar: IFAN, 1963.
Brooke, James. "Young Voice in Africa: Sports and Clean Living." *New York Times*, 7 September 1987, 8.
Burkina Faso, Government of. *Burkina Faso 1990: Country Presentation*. Paris: United Nations Conference on the Least Developed Countries, 1990.
———. *Code des investissements*. Ouagadougou: Government of Burkina Faso, 1988.
———. *Premier plan quinquénal de développement populaire, 1986–1990*. 2 vols. Ouagadougou: Ministère de la Planification et du Développement Populaire, 1986.
Cabral, Amilcar. *Unity and Struggle: Speeches and Writings*. New York: Monthly Review Press, 1979.
Capron, Jean. *Communatés villageoises Bwa: Mali-Haute-Volta*. Paris: Institut d'Ethnologie, 1973.
Clapham, Christopher. "Liberia: Recent History." In *Africa South of the Sahara 1995*. London: Europa Publications, 1994, 528–532.
Chambre de Commerce, d'Industrie et d'Artisanat du Burkina. *Symposium sur l'entreprise burkinabè: Bilan critique et perspectives d'avenir*. Vol. 2. *Rapport de synthèse du symposium, Août 1985*. Ouagadougou: Ministère du Commerce et de l'Approvisionnement du Peuple, 1986.
Club du Sahel. *Burkina Faso: Développement des cultures irriguées: Bilan critique, contraintes, propositions d'amélioration*. 2 vols. Ouagadougou: Comité Permanent Inter-Etat de Lutte contre la Sécheresse dans le Sahel, 1987.
Comité Interafricain d'Etudes Hydrauliques. *Etude des moyens de production de pluie provoquée: Expérimentation en Haute-Volta, année 1974–75–76*. Ouagadougou: Secrétariat Général du CIEH, 1977.
Conférence Nationale des CDR. *Première Conférence Nationale des CDR: Documents finaux*. Ouagadougou: SGN-CDR, 1986.
Conombo, Joseph Issoufou. *Souvenirs de guerre d'un "Tirailleur Sénégalais."* Paris: L'Harmattan, 1989.
Conseil National de la Révolution. *Discours d'orientation politique*. Ouagadougou: Ministère de l'Information, 1983.
———. *Programme Populaire de Développement, octobre 1984-décembre 1985: Bilan final*. Ouagadougou: Ministère de la Planification et du Développement Populaire, 1986.
Coordination du Front Populaire. *Mémorandum sur les evénements du 15 octobre 1987*. Ouagadougou: Front Populaire, 1988.
Coquery-Vidrovitch, Catherine, ed. *L'Afrique occidentale au temps des Français. Colonisateurs et colonisés, 1860–1960*. Paris: La Découverte, 1993.
Coulibaly, Sidiki, Joel Gregory, and Victor Piché. *Les Migrations voltaïques*. Vol. 1: *Importance et ambivalence de la migration voltaïque*. Ottawa: Centre Voltaïque de la Recherche Scientifique, Institut National de la Statistique et de la Démographie, 1980.
"Cross Cultural Study of Burkina Faso." Prepared for the Overseas Briefing Center Foreign Service Institute. U.S. Department of State, Washington, D.C., 1982. Mimeographed.
Crowder, Michael. *Colonial West Africa: Collected Essays*. London: Frank Cass, 1978.

Dahl, Robert, and Charles Lindblom. *Politics, Economics and Welfare*. New York: Harper and Brothers, 1953.
Davidson, Basil. *The Black Man's Burden: Africa and the Curse of the Nation-State*. New York: Times Books, 1992.
"Death of the More Naba." *West Africa*, 20 December 1982, 326.
Delafosse, Maurice. *Le Pays, les peuples, les langues, l'histoire, les civilisations du Haut-Sénégal-Niger*. 3 vols. Paris: Larose, 1923.
Deuxième Conférence Nationale des CDR des Universités. *Formation et emploi dans l'état démocratique et populaire—contribution de la jeunesse estudiantine révolutionnaire*. Pô: Secrétariat Général des CDR, 1987.
Devarajan, Shantayanan, and Jaime de Melo. "Evaluating Participation in African Monetary Unions: A Statistical Analysis of the CFA Zones." *World Development*, 15, (1987): 483–496.
_____. "Membership in the CFA Zone: Odyssean Journey or Trojan Horse?" Paper presented at the World Bank Conference on African Economic Issues, 1990.
Dia, Mamadou. "Development and Cultural Values in Sub-Saharan Africa." *Finance and Development*, December 1991, 10–13.
Diarra, Marthe Doka, and Marie Monimart. "Women and Sustained Development in the Sahel: An Analysis of National and Aid Agencies' Policies in Burkina Faso and Mali." OECD/Club du Sahel, Paris, 1989. Mimeographed.
Dicko, Hama. "Essai d'étude socio-économique du projet minier-ferrovière de Tambao. Les enjeux, les impacts, les répercussions du projet de développement des transports sur la base d'une voie ferrée Ouagadougou-Kaya, Dori, Tambao, Tin-Hrassan sur le développement socio-économique de la région et de la nation." Université de Ouagadougou, 1984. Mimeographed.
Dim Delobson, A. A. *L'Empire du Mogho Naba. Coutumes des Mossi de Haute-Volta*. Paris: Donat-Montchrestien, 1932.
Dubuch, Claude. "Langage du pouvoir, pouvoir du langage." *Politique Africaine*, 20 (December 1985): 44–53.
Duval, Maurice. *Un totalitarisme sans Etat: Essai d'anthropologie politique à partir d'un village burkinabè*. Paris: L'Harmattan, 1985.
Elbadawi, Ibrahim, and Nader Majd. *Fixed Parity of the Exchange Rate and Economic Performance in the CFA Zone: A Comparative Study*. Working Paper Series 830. Washington, D.C.: World Bank, Country Economics Department, 1992.
Englebert, Pierre. "Burkina Faso in Transition." *CSIS Africa Notes*, March 1990, 111.
_____. "Burkina Faso: Recent History." *Africa South of the Sahara 1995*. London: Europa Publications, 1994, 191–195.
_____. "Burkina Faso: Towards a Constitution and Economic Adjustment." In *Africa Contemporary Record 1989–90*. Vol. 22. New York: Holmes and Meyer, 1996, B9–B19.
_____. *La Révolution burkinabè*. Paris: L'Harmattan, 1986.
Etienne-Nugue, Jocelyne. *Artisanats traditionnels en Afrique Noire–Haute-Volta*. Dakar: Institut Culturel Africain, 1982.
Ewusi, Kodwo, S. K. Daapah, and Clement Atriadeke. *The Development and Resettlement of Oncho-freed Zones in the Volta River Basin of West Africa: A Report to the Food and Agriculture Organization*. Legon: Institute of Statistical, Social and Economic Research, University of Ghana, 1985.

Fafchamps, Marcel. "Solidarity Networks in Preindustrial Societies: Rational Peasants with a Moral Economy." *Economic Development and Cultural Change*, 41, 1 (October 1992): 147–174.
Faure, Yves-André. "Ouaga et Abidjan: Divorce à l'africaine? Les raisons contre la raison." *Politique Africaine*, 20 (December 1985): 78–86.
FESPACO. "Dixième Fespaco: 18 ans au service du cinéma africain." Ouagadougou: FESPACO, 1987.
_____. "FESPACO: 20ème anniversaire." Ouagadougou: FESPACO, 1989.
_____. *Festival Panafricain du Cinéma de Ouagadougou*. Paris: Présence Africaine, 1983.
Fieloux, Michèle. *Les Sentiers de la nuit: Les migrations rurales lobi de la Haute-Volta vers la Côte d'Ivoire*. Paris: ORSTOM, 1980.
Fiske, Alan P. *Structures of Social Life. The Four Elementary Forms of Human Relations: Communal Sharing, Authority Ranking, Equality Matching, Market Pricing*. New York: Free Press, 1991.
Frelastre, Georges. "La Politique agricole de la Haute-Volta est-elle à un tournant?" *Le Mois en Afrique*, 178–179 (October-November 1980): 66–73.
Front Populaire. *Guide d'organisation des Comités Révolutionaires*. Ouagadougou: Front Populaire, 1989.
_____. *Statuts et Programme d'Action*. Ouagadougou: Imprimerie Nationale, 1988.
Gonidec, Pierre-François. "Les Assemblées locales des territoires d'outre-mer." *Revue Juridique et Politique de l'Union Française*, 7 (1952): 328.
_____. *Les Systèmes politiques africains*. Paris: Librairie Générale de Droit et de Jurisprudence, 1978.
Gouilly, Alphonse. *L'Islam dans l'Afrique Occidentale Française*. Paris: Larose, 1952.
Granier, Roland, et al. "Disparités des revenus ville-campagne Côte d'Ivoire et Haute-Volta." Addis Ababa: Bureau International du Travail, Programme des Emplois et des Compétences Techniques pour l'Afrique, 1981.
Greenberg, Joseph H. "The Languages of Africa." *International Journal of American Linguistics*, 29 (January 1963): 8.
Gregory, J. W. *Underdevelopment, Dependency and Migration in Upper-Volta*. Ithaca, N.Y.: Cornell University Press, 1974.
Grobar, Lisa M. "Money and Macroeconomic Adjustment in the CFA Zone." California State University at Long Beach, 1994. Mimeographed.
Grootaert, Christiaan. "The Position of Migrants in the Informal Labor Markets in Côte d'Ivoire." World Bank, Development Research Department, Washington, D.C., 1987. Mimeographed.
Guillaume-Gentil, Anne. "Burkina Faso." *Marchés Tropicaux et Méditerranéens*, 47, 2364 (1 March 1991): 526–558.
Guillaumont, Patrick, and Sylviane Guillaumont, eds. *Stratégies de développement comparées: Zone Franc et hors Zone Franc*. Paris: Economica, 1988.
_____. *Zone Franc et développement Africain*. Paris: Economica, 1984.
Guion, Jean R. *Blaise Compaoré: Réalisme et intégrité. Portrait de l'homme de la "Rectification" au Burkina Faso*. Paris: Berger Levrault International, 1991.
Hargreaves, John D., ed. *France and West Africa: An Anthology of Historical Documents*. London: Macmillan, 1969.
_____. *Prelude to the Partition of West Africa*. New York: Macmillan, 1963.

_____. *West Africa: The Former French States*. Englewood Cliffs, N.J.: Prentice-Hall, 1967.
Harrison, Christopher. *France and Islam in West Africa, 1860–1960*. New York: Cambridge University Press, 1988.
Harrison Church, R. J. "Burkina Faso: Physical and Social Geography." In *Africa South of the Sahara 1995*. London: Europa Publications, 1994.
Harsch, Ernest. "Burkina Special Report: A Revolution Derailed." *Africa Report*, 33, 1 (1988): 36.
"Haute-Volta: Tension politique et sociale." *Afrique Contemporaine*, 124 (May-June 1982): 31.
Héritier, Françoise. "La Paix et la pluie, rapports d'autorité et rapport au sacré chez les Samo." *L'Homme*, 13, 3 (1973): 123–135.
Hill, Polly. *Migrant Cocoa Farmers of Southern Ghana: A Study in Rural Capitalism*. Cambridge: Cambridge University Press, 1970.
Holas, Bohumil. *Les Sénoufo*. Paris: Presses Universitaires de France, 1957.
Honohan, Patrick. *Monetary Cooperation in the CFA Zone*. Policy, Research, and External Affairs Working Paper Series 389. Washington, D.C.: World Bank, 1990.
_____. *Price and Monetary Convergence in Currency Unions: The Franc and Rand Zones*. Policy, Research, and External Affairs Working Paper Series 390. Washington, D.C.: World Bank, 1990.
Huntington, Samuel P. *Political Order in Changing Societies*. New Haven: Yale University Press, 1968.
Institut National de la Statistique et de la Démographie. *Annuaire statistique du Burkina Faso, 1989–1990*. Ouagadougou: Direction des Statistiques Générales, 1991.
_____. *Deuxième recensement général de la population du 10 au 20 décembre 1985: Principales données définitives*. Ouagadougou: Ministère du Plan et de la Coopération, n.d.
_____. *Enquête transport routier*. Ouagadougou: Ministère du Plan et de la Coopération, 1988.
_____. *Recensement général de la population, Burkina Faso 1985. Analyse des résultats définitifs*. Ouagadougou: Imprimerie de l'Institut National de la Statistique et de la Démographie, 1990.
_____. *Recensement général de la population, 1985: Structure par age et sexe des villages du Burkina Faso*. Ouagadougou: Ministère du Plan et de la Coopération, 1988.
Izard, Françoise, with Philippe Bonnefond and Marie d'Huart. *Bibliographie générale de la Haute-Volta, 1956–1965*. Paris: CNRS/CVRS, 1967.
Izard, Michel. *Gens du pouvoir, gens de la terre*. Cambridge: Cambridge University Press, 1985.
_____. *Introduction à l'histoire des royaumes mossi*. 2 vols. Paris: CNRS/CVRS, 1970.
_____. *L'Odyssée du pouvoir. Un royaume africain: Etat, société, destin individuel*. Paris: Editions de l'Ecole des Hautes Etudes en Sciences Sociales, 1992.
_____. *Le Yatenga précolonial. Un ancien royaume du Burkina*. Paris: Karthala, 1985.
Jaeger, William K. *Agricultural Mechanization: The Economics of Animal Draft Power in West Africa*. Boulder: Westview Press, 1986.
Kabeya-Muase, Charles. *Syndicalisme et démocratie en Afrique Noire: L'expérience du Burkina Faso, 1936–1988*. Paris: INADES/Karthala, 1989.
_____. "Un pouvoir des travailleurs peut-il être contre les syndicats?" *Politique Africaine*, 33 (1989): 50–58.

Kaboré, Victor G. *Organisation traditionelle et évolution politique des Mossi de Ouagadougou.* Ouagadougou: CVRS, 1966.

Kanse, Mathias. "Le CNR et les femmes: De la difficulté de libérer 'la moitié du ciel.' " *Politique Africaine,* 33 (1989): 66–72.

Kiéthéga, Jean-Baptiste. *L'Or de la Volta noire: Archéologie et histoire de l'exploitation traditionnelle, région de Poura, Haute-Volta.* Paris: Karthala, 1983.

Ki-Zerbo, Joseph. *Alfred Diban, premier chrétien de Haute-Volta.* Paris: Cerf, 1983.

Kouanda, Assimi. "Marabouts et missions catholiques au Burkina à l'époque coloniale (1900–1947)." Colloque d'Aix-en-Provence, n.d. Mimeographed.

Labazée, Pascal. "Discours et contrôle politique: Les avatars du sankarisme." *Politique Africaine,* 33 (1989): 11–26.

———. *Entreprises et entrepreneurs du Burkina Faso: Vers une lecture anthropologique de l'entreprise africaine.* Paris: Karthala, 1988.

———. "Hésitations et absence de dialogue: La voie étroite de la révolution au Burkina." *Le Monde Diplomatique,* 317 (February 1985): 12–13.

———. "Réorganisation économique et résistances sociales: La question des alliances au Burkina." *Politique Africaine,* 20 (December 1985): 10–28.

———. "Une nouvelle phase de la révolution au Burkina Faso." *Politique Africaine,* 24 (December 1986): 114–120.

———. "La Voie étroite de la révolution au Burkina Faso." *Le Monde Diplomatique* (February 1985): 12–13.

Lear, Aaron. *Burkina Faso.* Edgemont, Pa.: Chelsea House, 1986.

Lecaillon, Jacques, and Christian Morrisson. *Economic Policies and Agricultural Performance: The Case of Burkina Faso.* Paris: Development Centre of the OECD, 1985.

Ledange, Paul-Louis. "Une colonie nouvelle: La Haute-Volta." *La Revue Indigène,* 17 (1922): 133–136.

Le Moal, Guy. *Au Ghana avec les travailleurs voltaïques.* Ouagadougou: IFAN, 1965.

———. *Les Bobo: Nature et fonction des masques.* Paris: ORSTOM, 1980.

Lesselingue, Pierre. *Les Migrations des Mossi de Haute-Volta.* Ouagadougou: ORSTOM, 1974.

Lippens, Philippe. *La République de Haute-Volta.* Paris: Berger-Levrault, 1972.

Lockheed, Marlaine, Adriaan Verspoor, et al. *Improving Primary Education in Developing Countries.* Oxford: Oxford University Press, 1991.

Maiga, Mohamed. "Ouaga sur le qui-vive." *Afrique-Asie,* 301 (1 August 1983): 29.

Manning, Patrick. *Francophone Sub-Saharan Africa, 1880–1985.* Cambridge: Cambridge University Press, 1988.

Marchal, Jean-Yves. *Yatenga, nord Haute-Volta: La dynamique d'un espace rural soudano-sahelien.* Paris: ORSTOM, 1983.

Marie, Alain. "Politique urbaine: Une révolution au service de l'Etat." *Politique Africaine,* 33 (1989): 27–38.

Martens, Ludo. *Sankara, Compaoré et la révolution burkinabè.* Antwerp: EPO Dossier International, 1989.

Marx, Karl, and Friedrich Engels. *The Communist Manifesto.* New York: Monthly Review Press, 1964.

McFarland, Daniel Miles. *Historical Dictionary of Upper Volta.* Metuchen, N.J.: Scarecrow Press, 1978.

Ministère de l'Education Nationale. *Réforme de l'éducation: Dossier initial.* Ouagadougou: Ministère de l'Education Nationale et de la Culture, 1981.
"Mission to Ouagadougou." *World View,* 6, 3 (Summer 1993): 22–23.
Morgenthau, Ruth Schachter. *Political Parties in French-Speaking West Africa.* Oxford: Clarendon Press, 1964.
Mosley, Paul, Jane Harrigan, and John Toyle, eds. *Aid and Power: The World Bank and Policy Based on Lending.* London: Routledge, 1991.
"The Mossi and the Military." *West Africa,* 24 (January 1983): 18.
Nikiéma, Norbert. "La Situation linguistique en Haute-Volta: Travaux de recherche et d'application sur les langues nationales." UNESCO, Paris, 1980. Mimeographed.
Nol, Ned. "Le Mossi: La mission du Lt. Voulet." *Le Tour du Monde,* 33 (14 August 1897): 257.
"O Captain, Their Captain." *Economist,* 305, 7521 (24 October 1987): 52–54.
Organization for Economic Cooperation and Development (OECD). *Geographical Distribution of Financial Flows to Developing Countries: Disbursements, Commitments, Economic Indicators, 1985/1988.* Paris: OECD, 1990.
Otayek, René. "Burkina Faso: Between Feeble State and Total State, the Swing Continues." In Donal B. Cruise O'Brien, John Dunn, and Richard Rathbone, eds., *Contemporary West African States.* Cambridge: Cambridge University Press, 1989, 13–30.
———. "La Crise de la communauté musulmane de Haute-Volta: L'Islam voltaïque entre réformisme et tradition, autonomie et subordination." *Cahier d'Etudes Africaines,* 24, 3 (1984): 299–320.
———. "Rectification." *Politique Africaine,* 33 (1989): 2–10.
Ouédraogo, Bernard Lédéa. *Entraide villageoise et développement: Groupements paysans au Burkina Faso.* Paris: L'Harmattan, 1990.
Ouédraogo, Jean-Bernard. *Formation de la classe ouvrière en Afrique Noire: L'example du Burkina.* Paris: L'Harmattan, 1989.
Ouédraogo, Joseph, and André Prost. "La propriété foncière chez les Mossi." *Notes Africaines,* 38 (1948): 16–18.
Ovesen, Jan. *Ethnic Identification in the Voltaic Region: Problems of the Perception of "Tribe" and "Tribal Society."* Uppsala, Sweden: University of Uppsala, African Studies Programme, Department of Cultural Anthropology, 1987.
Pacere, Titinga Frédéric. *Ainsi on a assassiné tous les Mossi: Essai-témoignage.* Sherbrooke, Quebec: Editions Naaman, 1979.
———. *L'Avortement et la loi.* Ouagadougou: Imprimerie Nouvelle du Centre, 1983.
Pallier, Ginette. "Les Problèmes de développement dans les pays interieurs de l'Afrique occidentale: Contribution à l'étude du phénomène d'enclavement." Doctoral dissertation, Université de Bordeaux, 1984.
Peres de Arce, Diego. *Facteurs d'inflation en Côte d'Ivoire et en Haute-Volta.* Paris: Ministère des Relations Extérieures, Coopération et Développement, Sous-Direction des Etudes du Développement, 1983.
Person, Yves. "Les Syndicats en Afrique Noire." *Le Mois en Afrique* (172–173), April–May 1980: 22–45.
Porgo, Etienne. *Education en Haute-Volta: Analyse du secteur non formel.* Quebec: Université Laval, 1982.
Reardon, Thomas, Peter Matlon, and Christopher Delgado. "Coping with Household-Level Food Insecurity in Drought-Affected Areas of Burkina Faso." *World Development,* 16, 9 (1988): 1065–1074.

———. "Determinants and Effects of Income Diversification Amongst Farm Households in Burkina Faso." *Journal of Development Studies,* 28, 2 (January 1991): 264–296.

Reardon, Thomas Anthony; Taladidia Thiombiano; and Christopher Delgado. *La substitution des céréales locales par les céréales importées: La consommation alimentaire des ménages à Ouagadougou, Burkina Faso.* Ouagadougou: Université de Ouagadougou, Centre d'Etude de Documentation de Recherches Economiques et Sociales/IFPRI, 1988.

Robinson, Pearl T. "Grassroots Legitimation of Military Governance in Burkina Faso and Niger: The Core Contradictions." In Goran Hyden and Michael Bratton, eds., *Governance and Politics in Africa.* Boulder: Lynne Rienner Publishers, 1992, 143–165.

Rothchild, Donald, and Naomi Chazan, eds. *The Precarious Balance: State and Society in Africa.* Boulder: Westview Press, 1988.

Rouville, Cécile de. *Organisation sociale des Lobi: Une société bilinéaire du Burkina Faso et de Côte d'Ivoire.* Paris: L'Harmattan, 1989.

Russell, Sharon S., Karen Jacobsen, and William D. Stanley. *International Migration and Development in Sub-Saharan Africa.* Vol. 2: *Country Analyses.* Washington, D.C.: World Bank, 1990.

Sahn, David E., and Alexander Sarris. "States, Markets, and the Emergence of New Civil Institutions in Rural Africa." Paper presented at the conference on "State, Market, and Civil Institutions: New Theories, New Practices, and Their Implications for Rural Development," Ithaca, N.Y., December 13–14, 1991.

Sanders, John H., Joseph G. Nagy, and Sunder Ramaswami. "Developing New Agricultural Technologies for the Sahelian Countries: The Burkina Faso Case." *Economic Development and Cultural Change,* 39 (October 1990): 1–22.

Sangmpan, S. N. "Neither Soft nor Dead: The African State Is Alive and Well." *African Studies Review,* 36, 2 (September 1993): 73–94.

Sankara, Thomas. *Libération de la femme, une exigence du futur.* Ouagadougou: Editions du SGN-CDR, 1987.

———. "Oser inventer l'avenir." *La parole de Sankara.* Paris: L'Harmattan, 1991.

———. *Thomas Sankara Speaks: The Burkina Faso Revolution, 1983–1987.* Translated by Samantha Anderson. New York: Pathfinder 1988.

Sarraut, Albert. *La Mise en valeur des colonies françaises.* Paris: Payot & Cie, 1923.

Saul, Mahir. "Money and Land Tenure as Factors in Farm Size Differentiation in Burkina Faso." In R. E. Downs and S. P. Reyna, eds., *Land and Society in Contemporary Africa.* Hanover, N.H.: University Press of New England, 1988.

Savonnet-Guyot, Claudette. *Etat et sociétés au Burkina: Essai sur le politique africain.* Paris: Karthala, 1986.

———. "Le Prince et le Naaba." *Politique Africaine,* 20 (December 1985): 29–43.

Schildkrout, Enid. *People of the Zongo: The Transformation of Ethnic Identities in Ghana.* Cambridge: Cambridge University Press, 1978.

Secrétariat de la Zone Franc. *Rapport annuel.* Paris: Secrétariat de la Zone Franc, 1993.

Seni, Lazoumou. *La Lutte du Burkina contre la colonisation: Le cas de la région ouest, 1915–1916.* Ouagadougou: Imprimerie des Forces Armées Nationales, 1985.

Serageldin, Ismail, and June Taboroff, eds.. *Culture and Development in Africa.* Washington, D.C.: World Bank, 1994.

Sherman, Jacqueline R. *Grain Markets and the Marketing Behavior of Farmers: A Case Study of Manga, Upper Volta.* Ann Arbor: Center for Research on Economic Development, University of Michigan, 1984.

Skinner, Elliott P. *African Urban Life: The Transformation of Ouagadougou.* Princeton, N.J.: Princeton University Press, 1974.

———. "Christianity and Islam Among the Mossi." *American Anthropologist,* 60, 6 (December 1958): 1102–1119.

———. "The Diffusion of Islam in an African Society." *Annals of the New York Academy of Sciences,* 96 (20 January 1962): 659–667.

———. "Intergenerational Conflict Among the Mossi: Father and Son." *Journal of Conflict Resolution,* 5 (March 1961): 55–60.

———. "Labour Migration and Its Relationship to Socio-Cultural Change in Mossi Society." *Africa,* 30, 4 (October 1960): 375–401.

———. *The Mossi of Burkina Faso: Chiefs, Politicians and Soldiers.* Prospect Heights, Ill.: Waveland Press, 1989.

———. "The 'Paradox' of Rural Leadership: A Comment." *Journal of Modern African Studies,* 6, 2 (1970): 199–201.

———. "Processus de l'incorporation politique dans les sociétés africaines traditionelles: Le cas des Mossi." *Notes et Documentations Voltaïques,* 1, 4 (July-September 1968): 29–47.

———. "Sankara and the Burkinabé Revolution: Charisma and Power, Local and External Dimensions." *Journal of Modern African Studies,* 26, 3 (1988): 437–455.

———. "Strangers in West African Societies." *Africa,* 33, 4 (1963): 307–320.

———. "Trade and Markets Among the Mossi People." In Paul Bohannan and George Dalton, eds., *Markets in Africa.* Evanston, Ill.: Northwestern University Press, 1962, 237–278.

Sklar, Richard L. "Democracy in Africa." In Richard L. Sklar and C. S. Whitaker, eds., *African Politics and Problems in Development.* Boulder: Lynne Rienner Publishers, 1991, 249–264.

Somé, Valère D. *Thomas Sankara: L'espoir assassiné.* Paris: L'Harmattan, 1990.

Songre, Ambroise; Jean-Marie Sawadogo; and Georges Sanogoh. "Réalités et effets de l'émigration massive des Voltaïques dans le contexte de l'Afrique Occidentale." In Samir Amin, ed., *Modern Migrations in Western Africa.* London: Oxford University Press, 1974.

"Spécial Burkina Faso." *Marchés Tropicaux et Méditerranéens,* 2467 (19 February 1993): 466–500.

Speirs, Mike. "Agrarian Change and the Revolution in Burkina Faso." *African Affairs,* 90 (1991): 89–110.

Tallet, Bernard. "Le CNR face au monde rural: Le discours à l'épreuve des faits." *Politique Africaine,* 33 (1989): 39–49.

———. "Espaces ethniques et migrations: Comment gérer le mouvement?" *Politique Africaine,* 20 (December 1985): 65–77.

Tarrab, Gilbert, and Chris Coëne. *Femmes et pouvoirs au Burkina Faso.* Quebec: G. Vermette and L'Harmattan, 1989.

Thiombiano, Taladidia. *Une enclave industrielle: La société sucrière de Haute-Volta.* Dakar: Codesria, 1984.

Thompson, Virginia. *West Africa's Council of the Entente.* Ithaca, N.Y.: Cornell University Press, 1972.

Thompson, Virginia, and Richard Adloff. *French West Africa.* Stanford, Calif.: Stanford University Press, 1957.

Triaud, Jean-Louis. *Islam et sociétés soudanaises au Moyen Age. Etude historique.* Ouagadougou: CNRS/CVRS, 1973.

United Nations Development Program and the World Bank. *African Development Indicators*. New York: World Bank, 1992.

United States Department of State. *Country Report on Human Rights Practices for 1987*. Washington, D.C.: Government Printing Office, 1988.

Vallée, Olivier. *Le Prix de l'argent CFA. Heurs et malheurs de la Zone Franc.* Paris: Karthala, 1989.

van de Walle, Nicolas. "The Decline of the Franc Zone: Monetary Politics in Francophone Africa." *African Affairs*, 90 (1991): 383–405.

van der Gaag, Jacques, Elene Makonnen, and Pierre Englebert. *Trends in Social Indicators and Social Sector Financing.* World Bank Working Paper Series 662. Washington, D.C.: World Bank, 1991.

van Dijk, Meine Pieter. *Burkina Faso: Le secteur informel de Ouagadougou*. Paris: L'Harmattan, 1986.

Vengroff, Richard. *Upper Volta: Environmental Uncertainty and Livestock Production.* Lubbock: International Center for Arid and Semi-arid Land Studies, Texas Tech University, 1980.

Vinay, Bernard. *Zone Franc et coopération monétaire.* Paris: Ministère de la Coopération et du Développement, 1988.

Vizy, Marc. *La Zone Franc.* Paris: Centre des Hautes Etudes sur l'Afrique et l'Asie Modernes, 1989.

Whitaker, C. S. "Doctrines of Development and Precepts of the State: The World Bank and the Fifth Iteration of the African Case." In Richard L. Sklar and C. S. Whitaker, eds., *African Politics and Problems in Development.* Boulder: Lynne Rienner Publishers, 1991, 333–353.

_____. "A Dysrythmic Process of Political Change." In Richard L. Sklar and C. S. Whitaker, eds., *African Politics and Problems in Development.* Boulder: Lynne Rienner Publishers, 1991, 169–193.

Wigg, David. *And Then Forgot to Tell Us Why . . . : A Look at the Campaign Against River Blindness in West Africa.* Washington, D.C.: World Bank, 1993.

Wiseman, John A. *Democracy in Black Africa. Survival and Revival.* New York: Paragon House Publishers, 1990.

Woods, Dwayne. "Civil Society in Europe and Africa: Limiting State Power Through a Public Sphere." *African Studies Review*, 35, 2 (September 1992): 77–100.

World Bank. *Adjustment in Africa: Reforms, Results, and the Road Ahead.* Washington, D.C.: Oxford University Press for the World Bank, 1994.

_____. *Burkina Faso: Economic Memorandum.* Vol 1: *Main Report* and vol 2: *The Industrial Sector and Statistical Tables.* Washington, D.C.: World Bank, 1989.

_____. *Social Indicators of Development.* Baltimore: John Hopkins University Press, 1994.

_____. *Upper Volta: Investment in Human Resources.* Washington, D.C.: World Bank, 1983.

Yaméogo, Maurice, Sangoulé Lamizana, and Jean-Hubert Bazié. *Au Burkina Faso la parole est à tous.* Ouagadougou: Imprimerie Nationale, 1987.

Yarga, Larba. "La Fin de la troisième république voltaïque." *Le Mois en Afrique*, 182–183 (February-March 1981): 43–51.

_____. "Modernisation administrative et autorité traditionelle en Haute-Volta." Université de Nice, 1975. Mimeographed.

_____. "Les Prémices à l'avènement du Conseil National de la Révolution en Haute-Volta." *Le Mois en Afrique*, 213–214 (October-November 1983): 24–41.

_____. *Séparation et collaboration des pouvoirs dans le système constitutionnel voltaïque.* Nice: Université de Nice, Institut du Droit de la Paix et du Développement, 1983.

_____. "Le Tripartisme dans le droit public voltaïque." *Le Mois en Afrique,* 174–175 (June-July 1980): 114–129.

Younger, Stephen D., and Edouard G. Bonkoungou. "Burkina Faso: The Project Agro-Forestier—A Case Study of Agricultural Research and Extension." In Economic Development Institute of the World Bank, *Successful Development in Africa: Case Studies of Projects, Programs and Policies.* Washington, D.C.: World Bank, 1989, 11–26.

Younger, Stephen D., and Jean-Baptiste Zongo. "West Africa: The Onchocerciasis Control Program." In Economic Development Institute of the World Bank, *Successful Development in Africa: Case Studies of Projects, Programs and Policies.* Washington, D.C.: World Bank, 1989, 27–55.

Zachariah, K. C., and Julien Conde. *Migration in West Africa: Demographic Aspects.* Washington, D.C.: Oxford University Press for the World Bank, 1981.

Zachariah, K. C., and My T. Vu. *Africa Region Population Projections 1987–88.* Population and Human Resources Technical Note 87-19a. Washington, D.C.: World Bank, 1987.

Zahan, Dominique. "The Mossi Kingdoms." In Daryll Forde and P. M. Kaberry, eds., *West African Kingdoms in the Nineteenth Century.* London: Oxford University Press for the International African Institute, 1967, 152–178.

Ziegler, Jean, and Jean-Philippe Rapp. *Sankara: Un nouveau pouvoir africain.* Lausanne: Pierre-Marcel Favre/ABC, 1986.

Zolberg, Aristide R. *Creating Political Order: The Party-States of West Africa.* Chicago: Rand McNally, 1966.

Zoungrana, Cécile Marie. *Etude des mariages et des divorces dans la ville de Ouagadougou: 1980–1983.* Ouagadougou: Ministère de la Planification et du Développement Populaire, Institut National de la Statistique et de la Démographie, 1986.

# *About the Book and Author*

Poor even by the standards of West Africa and landlocked at the edge of the Sahel, Burkina Faso—the "Land of Men of Dignity"—has been plagued by political instability since independence from France in 1960. The country has suffered five military coups, the last of which cost the life of the outspoken and charismatic leader Thomas Sankara, who had waged war on poverty, corruption, and illiteracy.

Yet Burkina's growth was surprisingly strong during the 1980s, as it made the best of its meager assets in cotton, gold, and livestock. The country is also fortunate in its relative lack of ethnic conflict, and the several religions practiced—Islam, Christianity, and animism—peacefully coexist. Burkina has earned mixed reviews on the international stage, however, fighting two wars with Mali and supporting Taylor's rebels in the Liberian civil war.

In this textured introduction to Burkina Faso, Pierre Englebert highlights the historical and contemporary factors that account for the country's instability; considers the ethnic, religious, and social contours of the Burkinabè polity; examines in depth the country's economic policies and prospects; and analyzes Burkina's external relations. Looking toward the next millennium, he concludes by assessing the chances of the apparent recent drive toward a more democratic system.

Pierre Englebert is a doctoral candidate in political economy at the University of Southern California, a writer for the *Economist* Intelligence Unit and a former consultant at the *World Development Report* and the Western Africa Department of the World Bank. Since the publication of his first book, *La Révolution Burkinabè* (Paris: L'Harmattan, 1986), he has contributed articles to *Africa South of the Sahara, Africa Contemporary Record,* and *CSIS Africa Notes* and has coauthored several publications for the World Bank, including a chapter in *Adjustment in Africa* (1994). Englebert also holds degrees from the Université Libre de Bruxelles, where he is associate researcher and visiting lecturer at the Center for the Study of International and Strategic Relations, and from the Johns Hopkins University Nitze School of Advanced International Studies.

# Index

ADES. *See* Alliance pour la Démocratie et l'Emancipation Sociale
ADF. *See* Alliance pour la Démocratie et la Fédération
ADP. *See* Assemblée des Députés du Peuple
Afrique Occidentale Française (AOF), 2, 29, 35
Age demographics, 119
Agrarian and Land Reform Law (1985), 94
Agriculture
    cash crop production and, 86(table), 89–90
    food crop production and, 86(table), 87–89
    general economic conditions and, 83–87, 103
    livestock production and, 90–92
    as percentage of GDP, 83–84
    state strategies for, 92–96
    traditional methods of, 92
    women and, 140
AIDS (acquired immune deficiency syndrome), 133, 134
Airlines, 101
Algeria, 163
Alliance pour la Démocratie et la Fédération (ADF), 67
Alliance pour la Démocratie et l'Emancipation Sociale (ADES), 67
Alliance pour le Respect et la Défense de la Constitution (ARDC), 67
Animism, 127
AOF. *See* Afrique Occidentale Française
Arab Saharaoui Democratic Republic, 163
ARDC. *See* Alliance pour le Respect et la Défense de la Constitution

Arts programs, 141–143
Assemblée des Députés du Peuple (ADP), 68
Assimilationism, 137, 141
Association des Juristes Africains, 62
Association des Sunnites de Haute-Volta, 130
Authoritarianism
    first military regime and, 46–48
    political independence and, 44–45
    *See also* Dictatorship; Totalitarianism

BALIB. *See* Banque Arabe-Libyenne du Burkina
Banfora, 118
Banks, 100–101
Banque Arabe-Libyenne du Burkina (BALIB), 101, 163
Banque Centrale des Etats d'Afrique de l'Ouest (BCEAO), 100
Banque Internationale du Burkina (BIB), 101
Banque Internationale pour le Commerce, l'Industrie, et l'Agriculture du Burkina (BICIA-B), 101
Banque Nationale du Développement (BND), 101
Baogo, Yatenga Naba, 19
Bassolet, François, 35
Bayart, Jean-François, 71, 125
BCEAO. *See* Banque Centrale des Etats d'Afrique de l'Ouest
Benin, 155
Berthet, Max, 33
BIAO. *See* West African Banque Internationale de l'Afrique de l'Ouest

189

BIB. *See* Banque Internationale du Burkina
BICIA-B. *See* Banque Internationale pour le Commerce, l'Industrie, et l'Agriculture du Burkina
Binger, Louis, 15, 16
Binswanger, Hans, 138
Birifor, 16, 83
Birthrates, 117
Blackflies, 134
BND. *See* Banque Nationale du Développement
BNP. *See* French Banque Nationale de Paris
Bobo-Dioulasso, 118
Bognessan, Arsène Yé, 62, 63, 64, 68
Boisson, Pierre François, 27
Boliden International Mining, 97
Boni, Nazi, 28, 30, 31, 34–35, 48, 170
Brazzaville conference, 27, 36
Bulli, Yatenga Naba, 19
Burkinabè, etymology of word, 1, 7(n2)
Burkinabè-Libyan Friendship and Solidarity Society, 163
Burkina Faso
  climate of, 4, 5(map)
  colonization of, 18–32
  economic development of, 77–112
  foreign relations of, 149–164
  future challenges for, 172–174
  geographical features of, 4, 6
  independence movement in, 32–36
  location of, 2, 3(map)
  naming of, 1, 58
  political history of, 43–72
  precolonial history of, 9–18
  social geography of, 6
  society and culture in, 117–143
  success of, 168–169
  threats to, 170–172
  *See also* Upper Volta
Bwa, 17

Cabral, Amilcar, 1
Caisse de Stabilisation des Prix des Produits Agricoles, 93

Caisse Nationale de Crédit Agricole (CNCA), 101
Cash crop production, 86(table), 89–90
Castro, Fidel, 1
CATC. *See* Confédération Africaine des Travailleurs Croyants
Catholicism, 130–133, 135, 139
Catholic schools, 135
Cattle herding, 90–92
CBMP. *See* Comptoir Burkinabè des Métaux Précieux
CDRs. *See* Comités de Défense de la Revolution
CEAO. *See* Communauté Economique de l'Afrique de l'Ouest
Centres Regionaux de Production Agro-pastorale, 93
*Cercles* (administrative units), 20
Cereal production, 86(table), 87–89
CFD. *See* Coordination des Forces Démocratiques
CFDT. *See* Compagnie Française pour le Développement des Fibres Textiles
CFTC. *See* Confédération Française des Travailleurs Croyants
Chad, 162, 163
Chambre des Représentants, 68
*Chefferie* (chief system), 122, 123, 124, 125
Child mortality, 118–119, 133
Christianity
  contemporary influence of, 132–133
  evangelization of the Mossi and, 130–132
Christmas War, 154, 155
CISL. *See* Confédération Internationale des Syndicats Libres
Cissé, Kader, 63
Civil society
  political state vs., 69–72, 173
  weakening of, 72
  women in, 137–140
Clientelism, 70
Climate, 4, 5(map)
CMHV. *See* Communauté Musulmane de Haute-Volta

*Index*

CMRPN. *See* Comité Militaire pour le Redressement et le Progrès National
CNCA. *See* Caisse Nationale de Crédit Agricole
CNPP. *See* Convention Nationale des Patriotes Progressistes
CNPP/PSD. *See* Convention Nationale des Patriotes Progressistes/Parti Social Démocrate
CNR. *See* Conseil National de la Révolution
Coastal access, 2, 4
Cohen, Herman, 160
Colonization
  early history of, 20–23
  economic organization of, 23–27, 79–80
  education and, 134–135
  Fourth Republic and, 27–30
  French conquest and, 18–20
  independence movement and, 32–36
  migration and, 110–111
  resistance to, 22–23
  taxation and, 21, 22, 23, 79
COMITAM. *See* Compagnie Minière de Tambao
Comité Militaire pour le Redressement et le Progrès National (CMRPN), 52–53, 124
Comités de Défense de la Révolution (CDRs), 55, 58–59, 124–125
Comités Révolutionnaires (CRs), 63
Communal sharing, 81
Communauté Economique de l'Afrique de l'Ouest (CEAO), 151, 153, 154
Communauté Musulmane de Haute-Volta (CMHV), 130
*Communist Manifesto* (Marx and Engels), 125
Communist Party
  French, 28, 29
  political revolution and, 55–61
Compagnie Française pour le Développement des Fibres Textiles (CFDT), 89
Compagnie Minière de Tambao (COMITAM), 98

Compaoré, Blaise
  as current president of Burkina, 1, 71
  ethnicity and, 123
  foreign policy of, 156–160, 162, 163–164
  military coups of, 55, 61
  photos of, 67, 132
  political orientation of, 54, 171–172
  rectification process of, 61–69
  religious policy of, 133
Comptoir Burkinabè des Métaux Précieux (CBMP), 97
Confédération Africaine des Travailleurs Croyants (CATC), 45, 46
Confédération Française des Travailleurs Chrétiens (CFTC), 31
Confédération Internationale des Syndicats Libres (CISL), 45
Confédération Syndicale Voltaïque (CSV), 51
Confédération Syndicale Burkinabè (CSB), 59
Congo, Issoufou, 27
Conombo, Joseph, 30, 31, 51, 70, 121, 170
Conseil de l'Entente, 151, 154
Conseil de Salut du Peuple (CSP), 54–55, 124
Conseil National de la Révolution (CNR), 55–61, 93–95, 124–125, 140–141, 152–155
Conseil Révolutionnaire Economique et Social, 63
Construction industry, 99
Consultative Committee, 47–48
Consumer price index (CPI), 89, 94(figure), 104–105(table), 106
Consumption, 106–107
Convention Nationale des Patriotes Progressistes (CNPP), 65
Convention Nationale des Patriotes Progressistes/Parti Social Démocrate (CNPP/PSD), 67
Coordination des Forces Démocratiques (CFD), 67
Côte d'Ivoire
  colonial migration to, 111

economic growth of, 80
  foreign policy with, 150–151, 153
  spread of AIDS from, 134
Cotton production, 86(table), 89–90
Coulibaly, Sidiki, 112, 170
Coups, military
  assassination of Sankara in, 61
  of Conseil National de la Révolution, 55–56
  overthrow of Saye Zerbo by, 53–55
  susceptibility to, 170
CPI. *See* Consumer price index
Crowder, Michael, 35
Crozat, François, 19
CRs. *See* Comités Révolutionnaires
CSB. *See* Confédération Syndicale Burkinabè
CSP. *See* Conseil de Salut du Peuple
CSV. *See* Confédération Syndicale Voltaïque
Culture
  economic development and, 80–83
  ethnicity and, 119–126
  religion and, 127–133
  role of women in, 137–140
  of the state, 141–143
  *See also* Social organization
Currency
  devaluation of, 106
  in Mossi kingdoms, 14

*Daba* (agricultural tool), 92
Dabo, Boukary, 65
Damiba, Pierre Claver, 67
Davidson, Basil, 143
Debts, foreign, 107–108, 116(n86)
de Gaulle, Charles, 25, 27, 32, 36
Delgado, Christopher, 89
Demography
  general statistics, 118–119
  of religions, 127
Depression, economic, 25
Diabré, Zéphirin, 69
Diallo, Salif, 69
Dictatorship
  of Sangoulé Lamizana, 46–48
  *See also* Authoritarianism; Totalitarianism
Diendéré, Gilbert, 64, 171
Dim Delobson, A. A., 10, 16, 19, 127, 137
Diplomacy. *See* Foreign relations
*Discours d'orientation politique (DOP)*, 57
Disease control, 133–134
Doe, Samuel, 158
Domestic migration, 116(n97)
Dorange, Michel, 30
Droughts
  agricultural production and, 84
  livestock production and, 90, 91–92
Dulugu, Mogho Naba, 128

ECOMOG. *See* ECOWAS Monitoring Group
Economic Community of West African States (ECOWAS), 157, 158
Economic organization
  agriculture and, 83–96
  culture and, 80–83
  French colonization and, 23–25, 79–80
  industrial underdevelopment and, 98–100
  macroeconomic performance and policies and, 102–110
  map of, 78
  migration and, 110–112, 116(n97)
  mining and, 96–98
  of Mossi kingdoms, 14–16, 80–83
  poverty and, 110–112
  services sector and, 100–102
  successful implementation of, 168–169
  women and, 139–140
ECOWAS. *See* Economic Community of West African States
ECOWAS Monitoring Group (ECOMOG), 158
Education, 134–136
Electoral abstention, 71(table), 72
Emigrants, 111–112
Employment
  French colonization and, 23–25, 79–80
  of women, 139–140
  *See also* Economic organization; Labor

Energy production, 99
Engels, Friedrich, 125
Enhanced structural adjustment facility (ESAF), 110
Entente Voltaïque, 29
Environmental degradation, 96
Environmental management project, 95–96
Equality matching, 81, 83
ESAF. *See* Enhanced structural adjustment facility
Ethnicity
  interethnic conflict and, 125–126, 172
  language and, 119
  map illustrating, 120
  Mossi influence and, 119–123
  national politics and, 168, 172
  precolonial diversity in, 16–18
  state policies and, 123–125
European Community, 108
Excision, 139, 141
Exports
  from Mossi kingdoms, 14
  products used as, 107
Eyadéma, Gnassingbé, 153, 156, 172

Family agricultural units, 92
Fergusson, George Ekem, 19
FESPACO. *See* Festival Panafricain du Cinéma de Ouagadougou
Festival Panafricain du Cinéma de Ouagadougou (FESPACO), 102, 141–142
Films, 141–142
FIMATS. *See* Force d'Intervention du Ministère de l'Administration Territoriale et de la Sécurité
Financial services, 100–101
Fiske, Alan P., 81–82
Folklore, 142
Food crop production, 86(table), 87–89
Force d'Intervention du Ministère de l'Administration Territoriale et de la Sécurité (FIMATS), 61
Foreign aid, 108
Foreign relations
  with France, 160–162
  Liberian Civil War and, 158–160, 164
  with Libya, 162–164
  period of 1960–1983, 150–151
  period of 1983–1987, 151–155
  period of 1987–1994, 156–160
Foreign trade
  with Ghana, 155
  by Mossi kingdoms, 14–15
  structural deficit in, 107
Fournier, Albéric Auguste, 23
Fourth Republic
  of Burkina Faso, 66–69
  of France, 27–30
FP. *See* Front Populaire
FPV. *See* Front Progressiste Voltaïque
France
  diplomatic relations with, 152, 157, 160–162
  early colonization by, 20–23
  Fourth Republic of, 27–30
  political independence from, 32–36
  precolonial conquests by, 18–20
Franco-African summits, 161, 162, 163
Franco-Arab schools, 136
Franc Zone, 106, 115(n80), 169
French Banque Nationale de Paris (BNP), 101
French Communist Party, 28, 29
French Free Forces, 25, 27
Front du Refus, 50, 51
Front Populaire (FP), 61–66, 156
Front Progressiste Voltaïque (FPV), 51, 52, 59

Gambia, 159
GCB. *See* Groupe Communiste Burkinabè
GDP. *See* Gross domestic product; Groupe des Démocrates Progressistes
GDR. *See* Groupe des Démocrates Révolutionnaires
Geldof, Bob, 56
Gender demographics, 118–119
Geographical features, 4, 6
Ghana, 2, 111, 134, 151, 153, 155, 156–157
GNP. *See* Gross national product

Gold Coast. *See* Ghana
Gold mining, 96–97
Government
   agricultural intervention by, 92–96
   authoritarian, 44–46
   coups to overthrow, 53–56
   ethnic policies of, 123–125
   military regimes in, 46–48, 52–53
   national renewal, 49–50
   *See also* Republics; State
Gregory, Joel, 112
Gross domestic product (GDP), 103, 104–105(table), 106
Gross national product (GNP), 77, 83–84
Groundnut production, 86(table), 90
Groupe Communiste Burkinabè (GCB), 61
Groupe des Démocrates Progressistes (GDP), 65
Groupe des Démocrates Révolutionnaires (GDR), 65
Groupe Solidarité Voltaïque (GSV), 32
GSV. *See* Groupe Solidarité Voltaïque
Gueye, Lamine, 30
Guinea, 151
Guissou, Henri, 28, 31
Gur family, 6
Gurmanche kingdom, 12, 16, 83
Gurunsi, 11, 16

Hacquard, Monsignor, 130
Handicrafts fair, 102
Harmattan wind, 4
Haut-Sénégal-Niger, 20, 23
Health issues, 133–134
Herding practices, 90–92
Hesling, Edouard, 23, 138
History
   of early colonization, 20–23
   of Mossi kingdoms, 10–12
   political, 43–72
   or precolonial era, 10–16
Houphouët-Boigny, Félix, 27–30, 31, 33, 36, 150–151, 153, 170, 172

ICO. *See* Islamic Conference Organization

Illiteracy, 135
IMF. *See* International Monetary Fund
Immunizations, 134
Imperialism, 151, 152
Imports
   to Mossi kingdoms, 14
   products acquired as, 107
Incest, 139
Income distribution, 103, 104–105(table), 106
Indépendants d'Outre-Mer (IOM), 28, 35
Independence movement, 32–36
Independent National Patriotic Front of Liberia (INPFL), 158
*Indigénat* system, 20–21
Individual Persons and Family Law Act, 140
Industrial underdevelopment, 98–100
Infant mortality, 133
Inflation, 103, 104–105(table), 106, 169
Informal economy, 99–100
INPFL. *See* Independent National Patriotic Front of Liberia
International Monetary Fund (IMF), 109–110
Interstar Mining, 98
Investments, 106
IOM. *See* Indépendants d'Outre-Mer
Irrigation schemes, 84–85, 93
Islam
   contemporary influence of, 130
   penetration of, 127–129
Islamic Conference Organization (ICO), 150
Israel, 150, 152, 157
Izard, Michel, 10, 129, 138

*Jeune Afrique* (weekly), 154
John Paul II, Pope, 132(photo)
Johnson, Prince, 158

Kaboré, Boukary, 62, 66, 157
Kaboré, Gaston, 142, 143
Kaboré, Roch Marc Christian, 65, 68–69
Kadiogo, 118
*Karité* production, 86(table), 90

*Index*

Kaya, 118
Keita, Drissa, 154
Keita, Modibo, 34, 174
Kérékou, Mathieu, 155
Kibissi, 11
ki-Zerbo, Diban Simon Alfred, 130–131
ki-Zerbo, Joseph, 45, 47, 48, 50, 53, 67, 68, 70, 171
Kom I, Mogho Naba, 128
Kom II, Mogho Naba, 20, 27
Konaré, Alpha Oumar, 156
Koné, Begnon, 45
Koudougou, 118
Kougri, Mogho Naba, 33, 121
Kountché, Seyni, 151, 153
Kutu, Mogho Naba, 128

Labor
 agricultural, 83
 French colonization and, 23–25, 79–80
 migration and, 110–112, 116(n97)
 organized unions for, 45–46, 59
 strikes and, 71–72
 women and, 139–140
 *See also* Economic organization
Labor Laws Act (1962), 140
Lamizana, Sangoulé, 46–52, 70, 130, 133, 136, 150–151
Land ownership, 92
Land tenure, 92
Language, 119
Le Moal, Guy, 111
Liberia, 158–160, 173
Liberian Civil War, 158–160, 164
Libya, 152, 155, 160, 162–164
Life expectancy, 133
Ligue Patriotique pour le Développement (Lipad), 51, 54, 55, 59, 60
Limestone mining, 98
Lingani, Jean-Baptiste Boukary, 55, 64, 171
Lipad. *See* Ligue Patriotique pour le Développement
Lippens, Philippe, 34
Literacy rate, 135
Livestock production, 90–92
Louveau, Edmond Jean, 25, 27

Lumumba, Patrice, 1, 152

Macroeconomic performance and policies, 102–110
Maize production, 86(table), 87
Mali, 36, 111, 151, 153–156
Mande, 6
Manganese mining, 97–98
Manning, Patrick, 35
Manufacturing, 98–99, 100
Market pricing, 81, 82
Marriage
 migration and, 112
 in Mossi society, 137–138
Marx, Karl, 125
Marxism, 125
Matlon, Peter, 88–89
McIntire, John, 138
MDP. *See* Mouvement des Démocrates Progressistes
MDV. *See* Mouvement Démocratique Voltaïque
Men
 migration of, 112
 as percentage of population, 118–119
Mendès-France, Pierre, 30
Migration
 colonial, 110–112
 contemporary, 111–112
 domestic, 116(n97), 126
 spread of AIDS and, 134
Military
 colonial conscription into, 22, 23, 25, 27
 factionalism in, 171
 Liberian Civil War and, 158–160
 political coups of, 53–56, 61
 upheaval and politicization in, 53
Military regimes
 of Sangoulé Lamizana, 46–48
 of Saye Zerbo, 52–53
Milk production, 91
Millet production, 86(table), 87–88
Mineral resources, 96–98
Mining, 96–98, 100
Mitterrand, François, 29, 161, 162

MLN. *See* Mouvement de Libération Nationale
MNR. *See* Mouvement National pour le Renouveau
*Mogho naba* (Emperor), 12, 14, 15–16, 21, 121, 122, 127
Momoh, Joseph Saidu, 159
*Monde, Le* (newspaper), 158
Money
  devaluation of, 106
  in Mossi kingdoms, 14
Monogamy, 131, 139, 140
Monteil, Parfait-Louis, 18
Mortality rate, 117, 133
Mossi kingdoms
  contemporary influence of, 119–123
  decline of, 16
  economic organization of, 14–16, 81–82
  ethnic diversity and, 18, 172
  French restructuring of, 20–21
  historical foundations of, 10–12
  marriage in, 137–138
  religion in, 127
  sociopolitical organization of, 12–14
  women and, 137
Mossi plateau, 4
Mouragues, Albert Jean, 28, 30
Mouvement de Libération Nationale (MLN), 45
Mouvement Démocratique Voltaïque (MDV), 31
Mouvement des Démocrates Progressistes (MDP), 64
Mouvement des Pionniers, 59
Mouvement National pour le Renouveau (MNR), 49
Mouvement Populaire d'Evolution Africaine (MPEA), 30
Movies, 141–142
MPEA. *See* Mouvement Populaire d'Evolution Africaine
Muslims, 127–130, 136

*Naam* (force of God), 12
Naboho, Kanidoua, 69
*Nakombse* (men of power), 12–13, 121

Nakourour, Mount, 6
*Nanamse* (chiefs), 138
*Napogsyure* system, 138
National Culture Week, 102, 142
National Parks, 102
National Patriotic Front of Liberia (NPFL), 158
National renewal, 49–50
*Nesomba* (ministers), 13–14
NGOs. *See* Nongovernment organizations
Niennega, 137
Niger, 151, 153, 155, 158
Ninissi, 11
Nkrumah, Kwame, 152
Nongovernment organizations (NGOs), 141, 169, 174
NPFL. *See* National Patriotic Front of Liberia
Nyerere, Julius, 152

OAU. *See* Organization of African Unity
OCP. *See* Onchocerciasis Control Program
ODA. *See* Official Development Assistance
ODP/MT. *See* Organisation pour la Démocratie Populaire/Mouvement du Travail
OECD. *See* Organization for Economic Cooperation and Development
Office National des Céréales (OFNACER), 93
Official Development Assistance (ODA), 152, 162
OFNACER. *See* Office National des Céréales
OMR. *See* Organisation Militaire Révolutionnaire
Onchocerciasis, 85, 134
Onchocerciasis Control Program (OCP), 134
ORDs. *See* Organismes Régionaux de Développement
Organisation Militaire Révolutionnaire (OMR), 60
Organisation pour la Démocratie Populaire/Mouvement du Travail (ODP/MT), 63–66, 171

*Index*

Organisation Voltaïque des Syndicats Libres (OVSL), 49
Organismes Régionaux de Développement (ORDs), 93
Organization for Economic Cooperation and Development (OECD), 88, 141
Organization of African Unity (OAU), 56
Otayek, René, 56, 124
Ouagadougou
   as capital city, 2, 37(n11)
   economic growth of, 79
   as Mossi kingdom, 11
   population of, 118
Ouahigouya, 118
Ouédraogo, Ablassé, 69
Ouédraogo, Clément Oumarou, 63, 64, 65, 68, 171
Ouédraogo, Gérard Kango, 31, 48, 49, 50–51, 68, 70, 151, 171
Ouédraogo, Idrissa, 142
Ouédraogo, Jean-Baptiste, 54–55, 123, 162
Ouédraogo, Joseph, 34, 45, 46, 48, 49, 50–51, 70, 121, 171
Ouédraogo, Macaire, 50
Ouédraogo, Mamadou, 28, 30
Ouédraogo, Pierre, 61
Ouédraogo, Youssouf, 68, 69
Ouennam (supreme god), 127
Ouézzin Coulibaly, Daniel, 27, 31–32, 33, 45, 70
OVSL. *See* Organisation Voltaïque des Syndicats Libres

Pacere, Titinga Frédéric, 82
PAI. *See* Parti Africain pour l'Indépendance
Palestine Liberation Organization (PLO), 150, 152
Parks, National, 102
Parti Africain pour l'Indépendance (PAI), 55
Parti Communiste Révolutionnaire Voltaïque (PCRV), 55–56
Parti Démocratique Unifié (PDU), 31–32
Parti Démocratique Voltaïque (PDV), 31–32

Parti du Regroupement Africain (PRA), 32, 44
Parti du Travail Burkinabè (PTB), 68
Parti National Voltaïque (PNV), 34
Parti pour la Démocratie et le Progrès (PDP), 68
Parti Républicain pour la Liberté (PRL), 34–35
Parti Social pour l'Emancipation des Masses Africaines (PSEMA), 30–32
PCRV. *See* Parti Communiste Révolutionnaire Voltaïque
PDP. *See* Parti pour la Démocratie et le Progrès
PDU. *See* Parti Démocratique Unifié
PDV. *See* Parti Démocratique Voltaïque
Peace Corps, 160
Pelletier, Jacques, 162
Penne, Guy, 152, 161
People's Republic of China. *See* Taiwan
Pères Blancs (Order of White Fathers), 130, 134
Pétain, Philippe, 25
Peuls, 6
Piché, Victor, 112
PLO. *See* Palestine Liberation Organization
PNV. *See* Parti National Voltaïque
*Pogsyure* system, 138–139
Polisario Front, 162
Political organization
   contemporary fourth republic and, 66–69
   of French colonization, 20–21
   French fourth republic and, 27–30
   independence movement and, 32–36
   of Mossi kingdoms, 12–14, 21
   political parties and, 30–32
   rectification process and, 61–66
   revolution and, 55–61
   roots of instability in, 69–72
   threats to, 170–172
Political party system
   corruption of, 44–45
   militarization of, 53–55
Polygamy, 131, 139, 140

Population
  agricultural problems and, 85, 91, 92
  statistics on, 117–118
Pô rebellion, 155
Poverty, 110–112
PRA. *See* Parti du Regroupement Africain
Precipitation patterns
  agriculture and, 84, 89–90
  livestock production and, 90, 91–92
  overview of, 4, 5(map)
Precolonial era
  end of, 18–20
  ethnic diversity in, 16–18
  history of, 10–16
Prices
  agricultural, 93–95
  economic growth and, 103, 104–105(table), 106
Private schools, 135
PRL. *See* Parti Républicain pour la Liberté
Production, economic, 103, 104–105(table), 106
PSEMA. *See* Parti Social pour l'Emancipation des Masses Africaines
PTB. *See* Parti du Travail Burkinabè

Qadhafi, Muammar, 162, 163, 164
Qur'anic schools, 136

Radio, 141, 162
Railroads, 101–102
Rainfall. *See* Precipitation patterns
RAN. *See* Régie Abidjan Niger
Randau, Robert, 137
Rassemblement Démocratique Africain (RDA), 28–30, 31, 35, 44, 50, 170, 173
Rassemblement du Peuple Français (RPF), 30
Rawlings, Jerry, 56, 153, 155
RDA. *See* Rassemblement Démocratique Africain
Reardon, Thomas, 88
Rectification, 61–66, 156–160
Régie Abidjan Niger (RAN), 101
Régiment Inter-Armes d'Appui (RIA), 51

Religions
  animism, 127
  Christianity, 130–133
  demographics on, 127
  Islam, 127–130
Republics
  first (1960–1966), 43–48
  second (1970–1974), 48–50
  third (1978–1980), 50–52
  fourth (1991–present), 66–69
  *See also* Government
Resistance movement, 22–23
Revolution
  ideological confusion on, 57–58
  military coup initiating, 55–56
  political disorder and, 60–61
  rectification process of, 61–66
  totalitarianism and, 58–60
  women in, 140–141
RIA. *See* Régiment Inter-Armes d'Appui
Rice production, 86(table), 87–88
River blindness. *See* Onchocerciasis
Roussin, Michel, 157
RPF. *See* Rassemblement du Peuple Français
Rural Integrated Development Program, 92
Ryallé, 137

Saaga II, Mogho Naba, 27, 29(photo), 33
SAF. *See* Structural adjustment facility
Salaries. *See* Wages
*Samba Traoré* (film), 142
Sana, Mogho Naba, 37(n11)
Sanitation, 134
Sankara, Mariam, 62
Sankara, Thomas
  on AIDS, 134
  assassination of, 61, 62–63
  character sketch of, 56–57
  ethnic policies of, 123, 124–125
  foreign policies of, 151–155, 161, 162
  naming of Burkina Faso by, 1
  political revolution and, 53–55
  on women, 140
Savings, 106–107

*Index*

Savonnet-Guyot, Claudette, 13, 17, 119
Saygo, Guy Lamoussa, 63
SCFB. *See* Société des Chemins de Fer du Burkina
Scholarships, 135–136
Schools, 134–136
Senegal, 36
Sénégambie-Niger, 20
Senghor, Léopold Sedar, 28, 174
Services sector
   financial services and, 100–101
   growth of, 103
   tourism and, 102
   transportation services and, 101–102
Sesame seed production, 86(table), 90
Sessouma, Guillaume, 65
Shea nut production, 86(table), 90
Sherman, Jacqueline, 88
Sierra Leone, 159
Sighiri, Mogho Naba, 19, 20
Sigué, Vincent, 61
Sikasso plateau, 6
Skinner, Elliott, 7(n2), 15, 17, 121, 125, 131, 137
Slavery
   French colonization and, 21
   in Mossi kingdoms, 15
SMG. *See* Société des Mines de Guiro
SMIG (minimum guaranteed salary), 93, 94(figure)
SNEAHV. *See* Syndicat National des Enseignants Africains de Haute-Volta
SNEB. *See* Syndicat National des Enseignants du Burkina
Social geography, 6
Social organization
   culture of the state and, 141–143
   demography and, 118–119
   education and, 134–136
   ethnicity and, 119–126
   health issues and, 133–134
   language and, 119
   of Mossi kingdoms, 12–14
   population and, 117–118
   of precolonial villages, 17
   religion and, 127–133
   women and, 137–141
   *See also* Culture
Société de Recherches et d'Exploitation Minières du Burkina (SOREMIB), 96
Société des Chemins de Fer du Burkina (SCFB), 101
Société des Fibres Textiles (SOFITEX), 89, 93
Société des Mines de Guiro (SMG), 96
Société Minière Coréo-Burkinabè (SOMICOB), 96
Society. *See* Civil society; Social organization
Soeurs Blanches, 131, 134
SOFITEX. *See* Société des Fibres Textiles
Somalia, 173
Somé, Valère, 61
Somé Yorian, Gabriel, 53–54, 55, 70, 152, 171
SOMICOB. *See* Société Minière Coréo-Burkinabè
SOREMIB. *See* Société de Recherches et d'Exploitation Minières du Burkina
Sorghum production, 86(table), 87–88
Sourou Valley Rural Integrated Development Program, 84–85
South Africa, 152
State
   civil society vs., 69–72, 173
   culture of, 141–143
   ethnic policies of, 123–125
   *See also* Government
Strikes, labor, 71–72
Structural adjustment facility (SAF), 109–110
*Structures of Social Life* (Fiske), 81
Sudan, 23
Sugar cane production, 86(table), 90
Superior Council of the Armed Forces, 47
SUVESS. *See* Syndicat Unique Voltaïque des Enseignants Secondaire et du Supérieur
Syndicat National des Enseignants Africains de Haute-Volta (SNEAHV), 48, 51, 59

Syndicat National des Enseignants du Burkina (SNEB), 59
Syndicat Unique Voltaïque des Enseignants Secondaire et du Supérieur (SUVESS), 48–49

Taiwan, 150, 151, 157
Tallet, Bernard, 95
*Talse* (men of power), 13–14
Tauxier, Louis, 79
Taxation
  French colonization and, 21, 22, 23, 79
  peasant poll tax repeal, 95
Taylor, Charles, 158–160
Technical services, 100
Television, 141
Temperature, 4, 5(map)
Tenga (goddess of the earth), 127
*Tenga* (spiritual power), 12
*Tengbiise* (men of the land), 12, 14, 127
Tenkodogo kingdom, 12
Tertiary sector, 99–100, 103
Thévenoud, Père Blanc Joanny, 28, 131
*Tilai Yaaba* (film), 142
Togo, 153, 156, 158
Totalitarianism
  political movement toward, 58–60
  *See also* Authoritarianism; Dictatorship
Touré, Adama, 56
Touré, Ahmed Sékou, 45, 151
Touré, Amadou Toumani, 156
Touré, Soumane, 51, 55, 59
Tourism, 102
TPR. *See* Tribunaux Populaires de la Révolution
Trade. *See* Foreign trade
Trade unions, 45–46, 59
Transportation services, 100, 101–102
Traoré, Moussa, 153, 154, 156
Tribunaux Populaires de la Révolution (TPR), 59–60
Tuareg rebels, 157
Tunisia, 164

UCB. *See* Union des Communistes Burkinabè

UDPB. *See* Union des Démocrates et Patriotes du Burkina
UDSR. *See* Union Démocratique et Socialiste de la Résistance
UFB. *See* Union des Femmes du Burkina; Union of Burkinabè Women
UGTAN. *See* Union Générale des Travailleurs d'Afrique Noire
ULC. *See* Union des Luttes Communistes
UMOA. *See* Union Monétaire Ouest Africaine
UNAB. *See* Union Nationale des Anciens du Burkina
UNDD. *See* Union Nationale pour la Défense de la Démocratie
UNDP. *See* United Nations Development Program
UNI. *See* Union Nationale des Indépendants
Union Démocratique et Socialiste de la Résistance (UDSR), 29
Union des Communistes Burkinabè (UCB), 60
Union des Démocrates et Patriotes du Burkina (UDPB), 65
Union des Femmes du Burkina (UFB), 59
Union des Luttes Communistes (ULC), 55
Union Générale des Travailleurs d'Afrique Noire (UGTAN), 45
Union Monétaire Ouest Africaine (UMOA), 100, 108, 154, 174
Union Nationale des Anciens du Burkina (UNAB), 59
Union Nationale des Indépendants (UNI), 50
Union Nationale des Syndicats des Travailleurs de Haute-Volta (UNSTHV), 45
Union Nationale pour la Défense de la Démocratie (UNDD), 50, 52
Union of Burkinabè Women (UFB), 141
Union Progressiste Voltaïque (UPV), 50
Union Révolutionnaire des Banques (UREBA), 101
Unions, labor, 45–46, 59

# Index

Union Syndicale des Travailleurs Voltaïques (USTV), 45, 46
Union Voltaïque (UV), 28–30
United Nations Development Program (UNDP), 108
United States, 160, 163
University of Ouagadougou, 135, 136
UNSTHV. *See* Union Nationale des Syndicats des Travailleurs de Haute-Volta
Upper Volta
  colonial economy in, 79–80
  colonization of, 18–32
  creation of, 23–27
  Fourth Republic of France and, 27–30
  independence movement in, 32–36
  origin of name, 7(n3), 23
  partition of, 25, 26(map)
  political systems in, 43–72
  precolonial social systems in, 16–18
  renaming of, 1, 58
  revolution in, 55–61
  *See also* Burkina Faso
UPV. *See* Union Progressiste Voltaïque
UREBA. *See* Union Révolutionnaire des Banques
USTV. *See* Union Syndicale des Travailleurs Voltaïques
UV. *See* Union Voltaïque

Vaccinations, 134
Vegetation patterns, 4
Vichy Republic, 25, 27
Village structure, 17
Voice of the Voltaic Revolution radio station, 162
Vokouma, Prosper, 156, 160
Voltaic family, 6
Volta Rivers, 4, 7(n3)
Voter abstention, 71(table), 72

Wages
  for agricultural workers, 93–94
  French colonization and, 24
  *See also* Labor

Wealth
  Mossi distribution of, 15–16
  *See also* Economic organization
Weber, Max, 143
*Weend Kuni* (film), 142, 143(photo)
West African Banque Internationale de l'Afrique de l'Ouest (BIAO), 101
White Sisters. *See* Soeurs Blanches
Wobgo, Mogho Naba, 19–20
Women
  migration of, 112
  as percentage of population, 118–119
  in the revolution, 140–141
  societal role of, 137–140
Women's liberation, 140
Work. *See* Employment; Labor
World Bank, 83, 88, 95, 108, 109–110
World Food Program, 95
World Health Organization, 134
World War I, 22, 23
World War II, 25, 27

Yaméogo, Denis, 46
Yaméogo, Edouard, 46
Yaméogo, Herman, 50, 64, 66–67, 68, 69, 70, 171
Yaméogo, Maurice
  death of, 68
  ethnicity and, 121, 123–124
  foreign policy of, 150–151
  political rise of, 31, 33–35, 170, 171
  as president of Upper Volta, 44–46, 69–70
  restoration of political rights of, 66
Yarse, 128
Yatenga kingdom, 11–12, 117
*Yirs* (family farms), 16–17
Yougbaré, Dieudonné, 131

Zaire, 173
Zerbo, Saye, 51–54, 70, 171
Zinc mining, 97
Zinda Kaboré, Philippe, 27–28, 33, 121, 170
Zongo, Henri, 55, 64, 171
Zoungrana, Paul, 51–52, 131, 133